EMPIRES OF OIL

"If you read just one book on oil this year, let it be by Duncan Clarke, whose *Empires of Oil* is a stimulating mix of futurology, philosophy and *realpolitik,* with a dose of lateral thinking thrown in. Drawing on sources as diverse as Machiavelli and Gibbon, Niall Ferguson and Robert D. Kaplan, as well as lessons learnt from three decades as a leading industry adviser, Clarke challenges current assumptions about the future of oil. The result is a blast of fresh intellectual air that forces a radical reappraisal of the role of the commodity that shapes our lives, for better or for worse."

Michael Holman, former Africa Editor, Financial Times

"Having known about Duncan Clarke's intimate knowledge of the Asian and world upstream since the mid 1980s, it is no surprise to find in *Empires of Oil* a masterly rendition of our past, present and future in the global oil exploration and development game. State oil companies have seized many commanding heights, and private companies must adjust to new terrains. Clarke's *tour du monde* tackles these issues with originality, new thinking and an unrivalled command of the complex strategies involved."

Dr Fereidun Fesharaki, chairman & CEO,
Facts Global Energy Group

"*Empires of Oil* is not only the stuff of history but also reflects the titanic struggles going on here and now within the complexities of world oil geopolitics. Few are more qualified than Duncan Clarke to give us this erudite account of such a critical unfolding drama. A world class economist with extensive exposure to many empires of oil, his interaction with all the players in the global oil game over thirty years provides a striking vision from one of the best non-linear thinkers of our age."

Conrad Gerber, president, Petro-Logistics Ltd., Geneva

"There is no one in the world untouched by the oil industry, as Duncan Clarke's fascinating *Empires of Oil* shows, by unravelling the strengths and the weaknesses of the world's oil companies and examining their adversaries in depth, in an intriguing story of how, one brick at a time, the modern barbarians have sought to subdue it, not only in the Middle East, but with initiatives arising from East Asia and holding many lessons for African governments."

Moeletsi Mbeki, deputy chairman,
South African Institute of International Affairs

"A timely contribution to the debate on corporate oil as the world fast-forwards into an uncertain energy future. With characteristic iconoclasm, Clarke critically evaluates urgent challenges presented by geopolitical transformations, explaining why smart companies must change strategies, portfolio choice and tactics to survive. The old order based on large players exercising unbridled power around the globe is under siege from smaller, more nimble competitors, while state oil companies worldwide are redefining the rules of the global oil game, with resource nationalists and ethnic militia redrawing the oil map to restrict entry and reserve access for traditional players."

Barry Morgan, Africa correspondent, Upstream, *the international oil and gas newspaper*

"With a wealth of knowledge derived from decades inside the world oil industry working with governments, national oil companies and players of all corporate types, Duncan Clarke draws unique and pertinent scenarios for global energy based on deep understanding and enlightening insights from history. A book to read, with lessons for all involved in petroleum and significant reflections on humanity's oil and gas future."

John M. Albuquerque Forman,
former director ANP (Brazil Oil & Gas Regulatory Agency)

Praise for Duncan Clarke's previous book,
The Battle for Barrels

"*The Battle for Barrels* says we should take warnings of impending Armageddon with a pinch of salt ... a useful corrective to Strahan's argument that the end is nigh."

Guardian

"... such a welcome addition to what has largely been a one-sided debate outside of energy industry circles. Clarke, well known as a leading corporate strategist within the industry, is well equipped to provide a clearly reasoned and balanced response to the gloom and doom predicted by peak oilers. Clarke's book may be the best overview of the lobby written by either friend or foe ... an optimistic alternative to peak oil Cassandras."

Petroleum Africa

EMPIRES OF OIL
Corporate Oil in Barbarian Worlds

Duncan Clarke

P

PROFILE BOOKS

First published in Great Britain in 2007 by
Profile Books Ltd
3A Exmouth House
Pine Street
London EC1R OJH
www.profilebooks.com

Typeset in Times by MacGuru Ltd
info@macguru.org.uk
Printed and bound in Great Britain by
Clays, Bungay, Suffolk

A CIP catalogue record for this book is available from the British Library.

ISBN 978 1 84668 046 5

The paper this book is printed on is certified by the © 1996 Forest
Stewardship Council A.C. (FSC). It is ancient-forest friendly. The printer
holds FSC chain of custody SGS-COC-2061

FSC

Mixed Sources
Product group from well-managed
forests and other controlled sources

Cert no. SGS-COC-2061
www.fsc.org
© 1996 Forest Stewardship Council

Contents

For Charmaine

Acknowledgements

In writing this book I have drawn on over thirty years of encounters with a wide variety of people working in various capacities and at diverse levels in developing worlds both inside the global oil industry and without. Many knowledgeable individuals have, not always wittingly, helped to shape the ideas on which this treatise is based. The core notions come from reflection over the years on the history of oil and an experience of the societies in which my work in the industry has been conducted, over 100 in all, as well as from discussions with people from many walks of life. None of these people is responsible for any errors of commission or omission that may be found in this text.

The argument of this book is informed by the disciplines of economics, social science and grand strategy, as well as petroleum economics. Several fine historians have provided inspiration over many years, together with a range of writers, thinkers and individuals met by chance – some from disciplines or professions far beyond the realms of either economics or the oil industry.

There is a natural tendency to view the world only from our own perspective. When this viewpoint is supplemented by the findings of research, there is a temptation to conflate the outcome with objectivity. To counter this tendency, it may help to disclose some personal aspects that lie behind the thinking of this book.

The experience of growing up in landlocked central Africa

during turbulent times and then travelling widely across the developing world has provoked awareness of the dramatic sweep of historical events over the past half-century. Although it has always been clear that the fates of empires, nation states, governments and corporate oil are intertwined, it has taken time to distil the essence of the relationships between them. While these interfaces change continuously, it seems there is also something umbilical and timeless about them.

Some exposure in early days to the history, literature and mythology of ancient Greece and Rome, augmented by an interest in the dramas of the ancient world, has induced a sense of recurring paradigms. Historiography can appear to be timeless: the ancient world provides models that may illuminate our contemporary world.

Background inevitably shapes outlook. This has been true for all those involved in the oil game, including its historians. Where we sit often reflects where we might stand. From this point of view, nomadic life has proved an advantage. Travel to over 100 countries during the course of four decades has provided multiple prisms through which to view the divergent societies and diverse views that make up our contested worlds. This experience has provided a panorama to distil a multifaceted account of the drama of the modern oil game.

Central to this book is the unfashionable but historical notion of "barbarians", used in an analogy drawn from the ancient world. The thesis is likewise informed by a strong personal sense of what it means to be a barbarian in modern times. This follows from an upbringing in Zimbabwe (then Rhodesia), where my family was implanted in the late 19th century. Many members remain, though many more are now dispersed across the world. My own wanderings over the past 30 years and more were induced, one way or another, by Africa's rapid and uneven decolonisation, its acrimonious state conflicts within declining empires, experience of mandatory UN economic sanctions and exposure to its most circumscribed oil regimes. In the process my world became one of protracted civil war, several regime changes, subsequent political dramas akin to (and, for many, worse than) ethnic cleansing, and the implosion of social and economic order.

Much time, especially during the 1970s and 1980s, was spent working in all parts of Africa and most of the rest of the developing world, areas that were ruled by flawed and unstable regimes. All this has proved highly instructive for understanding the rivalries between the great powers and the struggles of nation states to keep above the shifting water line of survival. Most states were then allied to one great power or another, whether Sino-Soviet or Western (non-alignment being either a fiction or a fantasy). Guerrilla wars, rebellions and complex social conflicts abounded, and several continue.

Such experience provided exposure to many modes of subsistence existence. An understanding of what I have termed "barbarian" worlds was essential for survival. The impact of direct experience about the "Other" proved raw and profound. It clarified vexed issues of social identity, generated insights into the complexities of divided and ethnic-based societies, and encouraged an appreciation of others' perspectives and their conditions of survival.

For over a decade during the last years of the Cold War, I worked in Geneva. This provided rich opportunities to learn from observation, and offered encounters with many of the adversaries. There was direct experience inside the UN as an economist working on and in Africa, especially through advisory work in economics, development and finance for some of its principal agencies. Part of my brief was to analyse the economics of weak states (then called least developed countries) within the context of tense third-world geopolitics, complete with the impacts of economic sanctions on selected pariah states. This period involved encounters with rebels, some armed and dangerous, and even the practitioners of espionage, which in that environment was rife.

Discovery of the world oil industry followed from the time that I spent examining its history and corporate strategies within the much-divided and contested African continent. There political upheavals, coups, military regimes, one-party states and various brands of socialism were all the rage while, at the same time, Western interests were never far away. Rivals were often locked in long, hot, proxy wars.

Thereafter came almost a decade working in joint ventures for and later with Petroconsultants, then the leading consulting and technical firm in the world oil industry. This provided privileged vistas on the global exploration game and the opportunity to meet numerous corporate executives, individuals, companies and government players who were shaping the upstream oil industry around the world.

Over the past 20 years, as chairman and CEO of Global Pacific & Partners, there has been the privilege to encounter a multitude of movers and shakers in the hydrocarbons industry. Our private advisory firm is a global business active in all continents. It brings us regularly into direct contact with the top players in government and corporate oil, each with their unique interests, diverse bodies of knowledge, and widely varying histories and perspectives. Some of these individuals may recall our discussions, others will not. Over the years these encounters have included some state presidents, OPEC dignitaries, numerous oil ministers, a range of chief executives, multitudes of government oil officials, many senior exploration executives, and technical managers from virtually all the key companies in the western world – and many beyond, including those still shaping the world oil future.

All this has occurred across several decades of great oil industry transformation, corporate restructuring, portfolio adjustment and serial crises in world affairs. The representatives met from corporate oil have included all sorts – the irreverent, the uninitiated, the powerful, the knowledgeable and the flawed. On the other side have been many leading players from governments and national oil companies in Asia, Africa, Latin America, the Middle East and Russia, along with their counterparts in the Western world. These parties hold divergent worldviews, some antithetical to the models espoused in the West. Some may be classified under the rubric of "barbarians" as deployed here – though I use this term with much respect.

The terrain of barbarian worlds has long held a deep fascination, stemming from research into development, before "discovering" oil, conducted on plantations, among indentured contract workers, in peasant villages and in urban shanties in several

African countries. The economics of masters and servants was one such interest, and apposite to the Roman/Barbarian model. This experience involved wide exposure to theories and ideas that have been developed to interpret this complex social and political arena as well as, later, this shadow side of the global upstream business built in and around developing worlds. These domains, in which issues of survival are confronted with many, sometimes sophisticated, counter-strategies, are in complex ways connected to the world energy industry. Their fragmented *demi-mondes* and shadow economies manifest a deep sense of shared yet also divided interest. Such optics now shape important parts of the world oil game within an industry that is probably the most integrated of global cultures.

It has been my great fortune to have met in recent times Professor Peter R. Odell, an economist of distinction on global oil matters and one architect of the best analytical works on world oil, spanning a publishing period on the economics of oil from the 1960s. There is no one still active in this game with such a distinguished pedigree, achievement and longevity. Many cogent comments received from him, with much gratitude, have helped correct some initial errors and clarified the arguments made, all of which (flaws and faults inclusive) remain in no manner attributable to anyone other than the author.

Not all parties will concur with the perspectives expressed in this text, and hence it is accepted that some subjectivity has been unavoidable. The observations made on the oil game are not intended as sceptical or pessimistic. The former can be a valuable vantage point from which to judge complex oil games.

The latter may merely reflect discovery of the actual conditions often found.

I deeply appreciate the contribution of my publisher, Profile Books. Even before they had published my preceding book, *The Battle for Barrels*, in early 2007, Andrew Franklin (publisher and managing director) and Daniel Crewe (senior editor) committed to forging ahead with this new look at understanding the global oil game. The skilled team at Profile Books had much to do with the final outcome. Many thanks are due to them all. Anthony Haynes has once again worked on the shaping of this text, deploying a sharp insight into improving its format, with dedicated editing providing great clarity. Once again I have much appreciation to render for these sterling inputs.

Some of the core ideas in *Empires of Oil* were first articulated in our strategy briefings conducted around the world over the past two decades, and I have drawn in part on recent private research, not in the public domain, conducted over the years for Global Pacific & Partners. Many thanks as always are due to the fine team in our company, who over many years have built our worldwide industry relationships with corporate and government players in the international exploration business. They may not know it, but without them this work would not have been possible.

Duncan Clarke

May 2007

Introduction

A complex global oil game is afoot, one that has always been there and is now, more than ever, in the mind of the public at large. It is contested by many parties including corporate oil, nation states, governments and national oil companies. They have been joined in the game by a plethora of private and social interests, more or less organised, each either holding or claiming a stake in the world oil patrimony.

Changes within and around the oil industry form a vast dance from which the future of our societies will emerge. Understanding the pattern of this dance is crucial to understanding the world's economic, political and social futures. The task of this book is to provide new perspectives with which to interpret the dance. It will examine not only the American empire of oil, but also several competing, emerging empires – as well as what has been termed above the "barbarian" polities to be found within and beyond those empires. In attempting this task, we shall seek help from a number of intellectual guides. They include two famous thinkers who, though much revered, are not normally thought of as authorities on, but whose insights and seminal ideas might be relevant to, the world oil industry: Niccolò Machiavelli (1469–1527) and Edward Gibbon (1737–94).

The analysis here is informed throughout by the idea of empire. Through the course of history there have been numerous such entities, many without recourse to oil – though since its discovery, oil has become a key ingredient in their rise and fall. Today petroleum is central to the workings of the forces shaping grand strategy – to the rivalries between great powers, the pressures on the American empire, the Kremlin's energy strategy, the emergence of China and its voracious state oil players, India's renaissance, the traditional oil giants of the Middle East (Saudi Arabia, Iran, Iraq) and the mini-empires of oil now found in Africa, Latin America, Central Asia and elsewhere.

History, of course, is an essential guide to the interpretation of the past, present and future shape of the world oil industry. Yet it has distinct limits. We need a prism through which to view the divergent forces involved in global oil. This, as shown especially in Chapter 1, is where we may learn from that pre-eminent observer of human affairs and statecraft, Niccolò Machiavelli. By following and adopting his vision we will seek to expunge much subjectivity from this study and to achieve a more neutral, dispassionate, ruthlessly objective picture of the global oil game.

Part 1 uses the notions of Machiavelli as a lens through which to view many of the myths and models that have been advanced to explain the world oil game. Some of these are found to be instructive, others faulty, but all may have some use as foundations on which to build a better framework of understanding. The emphasis here is on reviewing how explanatory ideas have been used to construct an understanding of the complex puzzle that forms the world oil game and also on discovering whether a rather more overarching view of that puzzle, especially for the 21st century, can be constructed than has hitherto been achieved.

The concept of empire, with its many modern adaptations and revisions, is considered a template for the world oil industry: past, present and future. Whereas the images and analogies drawn from the Roman and subsequent empires were bounded in geography and sometimes shifting physical limits, our empires of oil find their roots in heartlands connected to worldwide oil domains controlled by the former.

The original oil empire was a personal dynasty, which John D. Rockefeller transmuted into a corporate behemoth that – along with private competitors, later aligned in combination as trusts – came to dominate the world oil market. From these beginnings evolved the famous Seven Sisters: Exxon (Esso), Shell, BP, Gulf, Texaco, Mobil and Socal (later Chevron). Together they provided the life-blood of the Anglo-American oil world and, in due course, of the American empire itself.

Part 1 also considers the so-called "oil curse" and the nexus of oil, social conflict and war. Both have shaped empire. Oil and war are often assumed to be linked, causally and inexorably. We examine their relationship and implications for the future, as well as complex interactions between corporate oil and world poverty, both much in the public's mind.

Part 2 focuses on the contemporary Western corporate oil game and the challenges provided by the rise of resource nationalism and of new empires of oil. Corporate oil, including what is often termed Big Oil (denoting the large super-majors) but which also refers to a long corporate tail of smaller companies drawn from around the world, faces a struggle to survive and succeed within a much changed, bounded world. There are new empires

3

of oil emerging that challenge the old order, with a paradigm shift already in process. The focus in particular is on how corporate oil's ambitions have clashed with the diverse, active – and growing – threats emanating from new state oil companies and their resource-nationalist governments. These competitors and threats are found operating inside complex developing-world environs controlled by barbarian polities. This will make for a rougher ride into the future for corporate oil.

The rise and erosion of Western empire was the dominant saga of the late 20th century. In this light, Part 2 identifies the empires of oil that have emerged since the end of the Cold War, now evident in fractious energy developments involving Russia, China, Iran, the Middle East and many new oil-rich "pretenders" across the world. The competing empires, states and allied national oil companies form a vivid moving collage in the upstream oil world (the term "upstream" denoting that segment of the industry concerned with exploration, discovery and hydrocarbons extraction). Are the changes observed in the oil world merely cyclical or do they entail now some more permanent shift towards a new paradigm?

No discourse on empire can afford to ignore the barbarians, the alter egos of the dominant oil powers. Much of the world's residual oil patrimony lies outside old empires in the developing world. An understanding of barbarian worlds is therefore critical to an assessment of contemporary and future oil landscapes. The barbarians with whom the Roman Empire was forced to engage comprised a number of distinctive peoples, each with its own social order yet many with a common interest in assaulting the empire. The same is true of a plethora of barbarians in the oil game today.

Part 3 highlights how these barbarian worlds have been a focus for forces found inside the old empires of oil: pressures coming from what are termed the "modern barbarians", informal or organised non-state entities opposed to corporate oil and some hostile to Western interests. It highlighted certain aspects of the barbarian world's struggle with corporate oil – in particular the role of modern ideologies (especially that of anti-globalism) and of certain non-governmental organisations (NGOs) – and the various strategies used by oil companies in self-defence. Many of these modern barbarians, some with long historical roots, are often declared enemies of corporate oil, Western oil interests and the hydrocarbons industry. We ask: Who are these opponents, how are they organised and what is their agenda, in what manner are they shaping corporate strategy and portfolios, and how might corporate oil respond to ensure survival?

Part 4 surveys the clash of cultures embroiled in the rival empires of oil. It offers a judgement on the global future of corporate oil and its Western anchor states, including the American empire of oil. It considers the question of corporate oil's survival and some implications for the West. There are new oil rivals in ascendance. They bring with them new difficulties for the companies that plough the exploration and development furrows, but also the potential at times for new partnerships. Has corporate oil yet found the best way forward? If not, what are the implications, and what would be the impacts for the West if the great oil game was "lost" – if indeed there is any end-game in oil to be expected?

Finally, it is worth adding a comment concerning Peak Oil.

Although the issue surfaces from time to time inside the world oil debate, it is not considered the key element explaining the global oil strategy taking place, nor can it account for the initiatives of great oil powers and their manoeuvres. It is not that the matter is unimportant. Rather it is because it has already received a complete treatment in my book *The Battle for Barrels*: so to cover the issue again would seem otiose.[1]

Part 1

GLOBAL OIL GAMES

There are many distinguished works on the world oil industry and its history. Many scholarly studies provide insights on the historic shifts shaping hydrocarbons, the players and personalities, the geopolitical forces and the commercial issues encountered in a host of countries endowed with crude and fossil fuels. Yet much in the contemporary world oil experience awaits explanation. We lack any overarching model, based on fundamental concepts, with which to explain the complexities of the global oil industry over time and to anticipate its future.

Here we consider the most enlightening visions and ideas yet to emerge from the literature generated from within and outside the industry. The emphasis throughout will be on the search for a new architecture of interpretation.

Though many distinguished works have emanated from the industry, they exhibit certain limitations. Typically, they abstract from wider grand strategy between world powers and societal conditions in developing countries, or restrict themselves to narrowly defined periods of time. Few of them seek to provide any coherent theory of the dynamics of world oil history. An exception is world oil market theory, its central weakness typically an abstraction from realpolitik and usual focus on variants of neoclassical economic models, many concerned with OPEC/ non-OPEC oil supply or producer and consumer OPEC/OECD market patterns of supply/demand dualism.

From beyond the industry, works have issued from a variety of traditions, including historiography, economics, military studies, the annals of diplomacy and statecraft, and even conspiratorial sources. Often they view oil in isolation from geopolitics and the shifting grand strategies that have shaped our oil-based civilisations – either that or they have been tied inexorably to just one of these anchors. Yet modern human societies are not wholly separate from their oil and energy environs. There is, indeed, much symbiosis between them. And these environments are not static but shift with the tides of history.

To bring together the disparate views that have been advanced, a vantage point is needed – one from which we may "see" at least the past and the present equally. We need to resist being seduced by the many partisan views on offer. Down that road lie subjectivity, preference, passion and distortion. The question, then, is where to look for a framework perspective capable of encompassing the complexities of the global oil game.

I

Machiavelli's prism

Having spent three decades in research and advisory practice on the global upstream industry – watching vast shifts in the oil game, the rise and fall of states, the comings and goings of oil companies, and paradigm changes in geopolitics – I have learnt to appreciate the virtues of ruthless objectivity. This, in turn, has led me to understand the importance of strategy in the global oil game.

This study suggests that the ways of "seeing" developed by Niccolò Machiavelli may be the best with which to observe the oil world. Our culture has, of course, given Machiavelli a bad press. The adjective "Machiavellian" carries negative connotations, having come to mean something akin to "devious, manipulative and untrustworthy". In fact, as author Michael White has demonstrated, Machiavelli has over half a millennium become the West's most misunderstood, yet most widely read, thinker.[2] Only the Bible in the Western world has probably been read by more people over a longer period of time. The reception of the works that Machiavelli produced from the labyrinthine world of the Medici court in Florence (complete with its feuds with rival statelets) has come over the years to span many cultures across the world. It is in those areas with which this book is most concerned – strategy, statecraft, diplomacy and historiography – that Machiavelli became most distinguished. The wisdom offered in such works as *The Prince* is not that of any evil, so-called "Machiavellian", adviser.[3] Rather it may be characterised as the fruits of honest, harsh, forceful, uncomfortable, provoking and acutely perceptive thinking.

Machiavelli lived in Florence at a time of dramatic cultural upheaval involving a complex tangle of power games, business interests and military ventures centred on the Medici family and court in the environs of competing statelets. Machiavelli's diagnoses, derived from his understanding of ancient and contemporary history, integrated an understanding of myriad interrelated concerns: Florentine society, its secrecy, power games, bureaucracy, periodic wars, fickle human behaviour, statecraft, rebellions, the workings of the Holy Roman Empire and the foundations of the political order.

A hallmark of Machiavelli's thought was its thoroughgoing realism. He sought to describe the world as it was, rather than as how some either imagined or hoped that it might be. He judged morality in relation to results rather than intentions, ideological correctness, or posturing. The key question at all times was how to maintain the state. For Machiavelli, understanding the state entailed an appreciation of diversity, transitions over time, the frailty of the social order and inner social weaknesses.[4] Piety was ruthlessly exposed as a mask for that well-known driver of strategy: self-interest.

Here the diagnosis draws on Machiavelli's vision both for its general virtues and for its particular applicability to the oil world. Let us now begin to survey that world through Machiavellian eyes. It is a world bathed in controversy and contentiousness. Large areas of the world's oilfields and frontier zones operate under pre-modern conditions. Vast differences exist between societies there and Western society, from which most of corporate oil emanates. The political regimes involved in today's oil game provide examples of virtually every type of political order that has littered the history of the past 500 years, including semi-feudal, pre-modern, dynastic, authoritarian, cruel and primitive. Our global matrices today transform themselves as this mix of states adjusts to the phenomenon of globalisation. It is inside the world of oil that this process has been most pronounced.

Corporate oil circulates within and around these shifting political and societal vortices, accompanied by a mixture of diplomatic initiative, hard-edged power, military force, social conflict and sometimes war. Secrecy and subterfuge have long been involved in the evolution of this hydrocarbon world. Despite the clamour for transparency and responsible governance, these phenomena will not disappear. As social formations come and go the disputatious search for control and ownership of oil continues. Regardless of spin produced by public relations departments, human frailties abound in the leaderships of both governments and many state companies. Diplomacy and statecraft continue to shape the world oil order and to influence the shifting of paradigms within which the players articulate strategies to assert their positions.

Institutions inside and surrounding the oil games are found to be inherently dynamic. In the petroleum world, little remains static for long. Rebellions contest the political orders that, often only temporarily, provide legitimacy to owners of the world's hydrocarbon interests. In the struggle for oil, order frequently gives way to chaos. Competition is embedded in the mix. Sometimes it takes extreme forms. Certainly the tradition and character of the nation state have not remained inviolate.

It is this world that the core of this book examines through the eyes of Machiavelli. It seeks – hopefully with the necessary tact and imagination – to treat Renaissance Florence, complete with its struggles with competing powers, as an analogy for the dynamics of the modern oil world. Yet for all his insight and rigour, Machiavelli provides us only with a foundation. To fully understand our oil world, we need to supplement our (in the non-pejorative sense) Machiavellian structure of understanding, founded upon the notion of the state, with another master concept – that of empire.

We may apply this concept to the oil world in at least three ways. First, there is the Anglo-American variant of empire, much challenged in the past and even more so today, despite its residual dominance. Second, there are the emergent geopolitical empires in oil and great-power politics now challenging the American empire and its far-flung energy domains. Lastly, and rather more tangentially, we may think of the dominant oil companies as empires in themselves. Corporate oil writ large, especially Big Oil, may likewise be seen as such a beast. Much of the analysis below may be thought of as a reflection of the interplay between these three types of empire.

One work of seminal importance in our understanding of empire is Edward Gibbon's *The Decline and Fall of the Roman Empire*, originally published in six volumes during the period 1776–88; its significance is considered in Chapter 2. The import here of the history of Rome is that it reminds us that, in placing the concept of empire at the centre of this analysis, we are in effect committed to employing one further concept – that of the barbarians. Ever since Gibbon it has been evident that we need to incorporate into our understanding of the Roman Empire – and of all those empires analogous to it, including empires of oil – an understanding of those complex groups on the edges. Their destiny was to contest its primacy and eat away its heart, destabilising its edifice of hegemony in the process.[5]

It is a central theme here that the binary opposition of "empire" and "barbarian" is especially useful for illuminating the contemporary oil world. Just as the concept of empire may be applied in a number of ways, so we may find many analogies of Gibbon's "other": the barbarians. Such a model allows us a means for understanding the threats to the oil industry in the 21st century. Here we must emphasise, for the avoidance of all doubt, that the term "barbarian" is used in this study without prejudice. Indeed, in terms of origins, history and analytic outlook, the author recognises a self-reflecting barbarian mindset.

The Machiavellian perspective, the theory of empire and the conception of new forces challenging the oil industry seen as modern barbarians – these are the foundation stones on which this book constructs an understanding of the modern oil world. From this framework we can generate a narrative with which to explain the origins, character and partial erosion of the empire of oil in the Western world. Provided that the diversity of both corporate oil and its barbarian foes are properly delineated, this narrative allows us too to intuit something of the future for world oil.

2

Myths and models

There is no shortage of interpretations of the shape, history, and future of world oil. They reflect a variety of interests, ideologies, and theories, sometimes in contradictory manner. We need to untangle these interpretations, select the best, most accurate, ideas, and build a more profound model. Here we look in turn at the perspectives of historiography, economics and geopolitics.

＊

There have been numerous historical studies of the hydrocarbons business, variously focused on governments, companies, personalities (from John D. Rockefeller to Lee Raymond, once of ExxonMobil), oil crises and oil-related wars. Perhaps the most prolific and enduring writer in the global oil world has been Peter R. Odell. In a significant, widely published and much translated book, *Oil and World Power*, and in even earlier works, Odell brought many of these structural issues to the fore.[6] This doyen of writers on the world hydrocarbons game has covered an enormous spectrum over the years: the geography of oil, western

Europe's gas and resources, world political relationships to oil, non-OPEC supplies, oil and energy in developing economies, reserve estimates both in East and West, oil and energy crises, energy and environment, regional energy supplies, OPEC and its role, the commanding heights of oil and global oil markets. Odell's work dealt with a wide range of then contemporary themes, some the focus of others: on the international oil system, the control of oil, the Seven Sisters and oil barons. This collection of deep analytic insights stands out as the benchmark in evaluations of world oil markets. Instructive as these studies have been, most were of a 1970s and 1980s vintage: since then, much has changed in world power balances and in the global oil game played in the first decade of the 21st century.

In surveying recent contributions, we should equally note the widely recognised masterpiece of epic storytelling and world oil historiography: Daniel Yergin's *The Prize*.[7] Published in 1991, it has become a standard text on the subject, built in part on the earlier analytical works emanating from the world of oil economics. It provides a perceptive chronicle of Hydrocarbon Man and, in particular, an intellectual bridge between oil and diplomacy. Yergin demonstrates with finesse how in the 20th century oil became the key to mastery for the great powers and the focus of great conflict. He explores oil's alliance with modern economies and its interconnections with national strategies and global politics. *The Prize* shows how our global hydrocarbon society has emerged from the flux of world history as powers have been made and broken on the back of oil wealth. In this magnum opus are revealed the large impersonal forces of markets and technologies with and against which the great personalities of oil politics and business have worked. Yergin traces too the emergence of some of the enemies and threats to this order.

With Yergin as a guide, along with Odell and others, we can trace the story of world oil through the 20th century. What might be termed Rockefeller's personal empire of oil gave way in time to an American combination of oil. The Great Game was in essence local at the outset. War, peace and a build-up of threats saw the formation of oil-based trusts in America. This attracted competitors both in America and beyond – notably the further growth of Russian oil and in parallel Royal Dutch. The 20th century witnessed many deals that at the time were hailed as "deals of the century", only for each to be surpassed in due course by even bigger transactions. Corporate oil collusion – a sort of imperfect precursor to the mergers and acquisitions of the late 20th century – emerged from this battle for survival against a background of wars and the collapse of Imperial Russia.

In time new threats grew in Persia under its imperial tutelage in the form of Anglo-Persian (subsequently BP). Interstate competition for oil became the common business of colliding empires. Churchill's intervention, converting the British navy from coal to oil, shaped the conditions for the oil future. Oil was to play a leading role in the grand struggle for Europe. Companies vied against each other for government preference within imperial domains, illustrating the way in which corporate oil's survival had become in part a matter of competitive preference.

The discovery of huge new oil reserves in the Middle East – except in Iran which had established hydrocarbons – transformed the scale and geography of the world oil game. All parties wanted a piece of this new Aladdin's cave. The Red Line Agreement by which competing corporate concerns carved out their sanctuaries within and around Mesopotamia was a precursor in the private sector to some of the asset acquisition moves associated with resource nationalism today. The 21st century equivalent might be the étatiste lunge for oil control found in Venezuela under Hugo Chavez and Russia under Vladimir Putin.

There followed a reshaped era in the relations between companies and nation states. Players such as Mexico emerged more significantly; Venezuela, building on efforts from the 1920s, began to enter the supply scene; and conflict with the Bolsheviks rearranged the Eurasian chessboard. Market surpluses of crude led to the redesign of corporate strategies and brought governments into play in the interests of stabilisation. Rising nationalism found expression in the distant oilfields that lay beyond the homelands of the Anglo-American empire. New deals had to be struck and contracts renegotiated as Arab sheikhdoms began to force the pace. These developments too resonate in our modern oil world as companies once again are required to rethink their strategies, renegotiate contracts, accept tougher terms and reposition corporate portfolios in oil and gas.

In the Second World War, the control of oil became a vital pivot. Japan sought hegemony over the South-East Asian oil heartland. In determining its war strategy, the Third Reich prioritised self-sufficiency in oil, derived from coal and acquisition targets in the Caucasus. A wave of fear swept across the Allied powers. America appointed its Petroleum Co-ordinator for National Defence – in effect an "oil tsar" – to mobilise supply. This was long before the Organisation of the Petroleum Exporting Countries (OPEC), cartels and quotas entered the game. The connections between oil and war intensified. There are some parallels between the situation in the Second World War and the scene today in concerns over Iraq, for example, and in global angst associated with the fear of scarcity and energy security that has resurfaced around the world.[8]

The post-war petroleum order assumed a new shape. America faced dependence on foreign oil, especially from the Middle East, and challenges from Arab and Soviet interests. One by one, nation states cut new terms and contracts with corporate oil against a background of coups and attempts to establish vassal state dependencies. World crises – notably Suez – contributed to a redrawing of the world oil map as concern over the security of supply moved up the West's geopolitical agenda. The process of decolonisation brought a plethora of competitive interests to the table, new relationships within the old empires and fragmentation in the world oil order. Now surpluses commanded the attention of government. The oil-exporting countries, led by Venezuela and including some newly independent states such as Nigeria and Libya, combined to form OPEC, a cartel in ambition that in time led to new frontiers opening in the non-OPEC world. In particular, zones within North Africa and the Sahara (Algeria inside OPEC), Asia (Indonesia an OPEC member) and Latin America (Ecuador also inside OPEC) emerged as a counter to OPEC's tightening grip. These developments related to the stakes in oil security, creating a new cartography of oil geopolitics.

As the battle for world mastery over hydrocarbons intensified in the second half of the 20th century, corporate control began to erode. A slew of countries assumed greater power in the oil world, especially after 1973. Even the development of the Alaskan and North Sea mega-fields did not ultimately halt this trend and insulate the Western world from wider dependency. The weapon of oil sanctions became increasingly important in the struggles over the control and management of Middle East supply. The power of OPEC, often oscillating with the market in down cycles, has nonetheless now gradually increased as the world economy's reliance on oil deepened. So arose, and continue, the worries over corporate oil's access to large-scale reserves and low-cost supply.

In the 1970s and 1980s, the geopolitical strategies pursued by America and the fading empires were closely entwined with the oil game. Great price shocks realigned the chessboard. Conflicts involving states such as Iran, which were hostile to America after 1979, intensified. The commoditisation of oil created difficulties in cartel practice. The dramatic plunge in the price of crude in 1986 forced the industry into restructuring, a process that continued into the 21st century. The accompanying takeovers have reshaped corporate oil in a process that will no doubt still operate to reconfigure industry patterns. Serial crises involving war and invasion in Iraq, Iran and Kuwait have brought a new set of problems for strategy and corporate portfolio management. The establishment of state players over the past few decades, and the later thrust from national oil companies seeking overseas portfolios worldwide from the end of the last century, have contributed to the radical transformation of the oil game.

This brief account is but a synopsis of earlier works and Yergin's own very much thicker description of the history of the oil game. *The Prize* provides a detailed chronicle: its main text runs to nearly 800 pages and is supported by maps and evidence found in dozens of pages of endnotes and references. In *Empires of Oil*, in contrast, the methodology is based on revealing, and applying to the present and future, the skeletal bones of discovered oil historiography, as well as elements of past-revealed concepts, theories and paradigms inside a new grand strategy model of world oil competition. This book seeks to complement earlier treatments with an approach that builds on established concepts and adds new insights to provide a rounder, more complete contemporary picture.

※

In the many profound works on history and international affairs that may illumine the shape of the future, Europe often plays a pivotal role – not least because of its importance in the shaping of world-scale empires and energy markets. The EU now stands as a residual semi-unified collection of such empires from the past with ambitions for a single energy market. In *Paradise and Power*, Robert Kagan has argued that, following the end of the Cold War – fought largely over the future of the old continent – Europe and America have diverged sharply in interest and alignment.[9] Some aspects of this transatlantic divergence have had an impact on the oil game. Shell, BP and ENI, for instance, have long pursued their own agenda. Some European companies (including Total) have, for example, discounted American edicts on sanctions in countries such as Iran, Myanmar and Sudan. Their upstream strategies reveal differences in vision and positioning. Centuries of colonial history and empire have left their marks on the strategies, and indeed the guile, of many European players. Their world view differs from America's.

Europe's empires have long since receded. Across vast swathes of the earth, Europe ceded its colonial domains in a surprisingly short period of time. The oil industry cannot now expect to have the state by its side in all the zones in which it operates. It seems improbable that governments will want to wage wars throughout the world on corporate oil's behalf. To that extent, the old empires have partly abandoned their children. Though weakened, however, Europe's influence and power across the globe survives in different forms.

In *The Breaking of Nations*, Robert Cooper has provided a vision of the new century.[10] The treatise points to the breakdown, following the dissolution of empires, of nation states and to the pre-modern world's reassertion of its presence. Corporate oil must now learn to deal with greater complexities. State fragility has already had profound implications across the oil landscape. It is as if a new *terra nullius* has established itself in our world and oil geography – one owned or at least claimed by many, different barbarians.

Any new imperialism will face deeper challenges than before and will probably need to be Spartan rather than Athenian in nature. This new optic makes for deeper uncertainty – if not pure anarchy, as envisaged by Robert Kaplan in his book, *The Coming Anarchy* – and increased global security concerns for oil.[11] Fragmentation produces multiple, much contested orders.

New dilemmas are inevitable. Any empire of oil will be severely constrained, as will the strategic and portfolio management of corporate oil.

Though we are far from all-knowing about the revised hydrocarbon domains, much is now known: from texts and from experience. From this vantage point we can at least peer into the future. In this study, however, we seek to go further than that. We ask whether we can formulate a new paradigm robust enough to stand the test of time and to provide the necessary explanatory power for interpreting the shifting collage of strategy, industry processes and global events that forms and shapes the hydrocarbon world. For all its rich content, historiography on its own is not enough for this purpose. We need something more, at least recognition of deep forces that drive the upstream paradigm.

＊

Here it is useful to consider the contribution of economics. The application of economic theory to the oil industry has usually focused on models of markets, supply and demand, the competing interests of consumers and producers, the impact of monopoly and of imperfect competition (cartels and oligopolies), short-term and long-term trends and cycles, the impact of the so-called "oil curse" on certain blighted nations and contemporary manifestations of resource nationalism. This, to a large extent, defines much of the literature within the discipline applied to oil.

Odell's prolific writings have, however, covered new and extensive ground. His recent outstanding work, *Why Carbon Fuels Will Dominate the 21st Century's Global Energy Economy*, provides a compelling analysis of the energy equations expected to apply over the century ahead, shaped by the drivers of global oil supply and demand.[12] Odell's analysis is based broadly on neoclassical theory and approaches tempered with strong doses of realism. It is an essential component of any profound understanding of future oil worlds.

This work is the culmination of much previous scholarship, dealing with the principal economic components of the game – the economic geography of oil, crude oil and world power, natural gas/oil substitution, the key industry forms and markets, non-OPEC supplies, third-world energy, oil crises, price shocks and the role of OPEC, and the shape of the oil industry. Much that can be learned from this record of achievement can fit convincingly into an economics of oil, allied to the theory of empire and its connections with oil and strategy.[13] But in oil it is not just pure economics that prevail, as Odell shows in many works. There is more, and much is woven in and around the interactions of oil economics and its paradigm to consider.

Another thinker who has shed new light on the economics of oil is Leonardo Maugeri. In *The Age of Oil* he combines an understanding of oil history, the upstream industry, corporate strategy and insider knowledge of the global game.[14] Maugeri is concerned to demythologise oil in a world in which it is popularly "identified with wars, greed, and unspeakable power plays orchestrated by transnational elites engaged in schemes worthy of spy novels".[15] He is particularly sensitive to the trap of catastrophism and the Peak Oil mythologies as a model found wanting, and is keen to test recurrent myths against the historical record.

Maugeri's economics deploys a theory of repetitive structural cycles. He argues that in the perception of the oil game, certain fears recur – the fear of running out of oil, for example, or fear about security of supply. Often they are related to global quests for the control of reserves. According to Maugeri, the dominant experience has been a history of oversupply and low prices. This raises the question of whether the existing phase of high crude prices and security premiums will prove short-lived and subject to future cyclical adjustment.

For Maugeri the contemporary manifestation of oil catastrophism is the dual fear that oil production will soon diminish as reserves become depleted and that Islamic terrorists will gain control over the supply of oil. Maugeri is phlegmatic about this catastrophism: "Nothing we experience today is a major departure from the historical cycles of the oil market."[16] Even if Islamic terrorists were to gain control of an oil-rich country, they would still be subject to the laws of economics. Thus no wolf at the door is envisaged. An important issue is whether the old empires of oil will be re-established, if and as crude prices deflate from new highs, or if they might not be able to be so reconstructed.

The Age of Oil identifies specific periods that have shaped imperial energy strategy, the linkages between national security and oil, and the processes of empire displacement (with an examination of America's umbilical relationship with Saudi Arabia). There is much substantiation of the central nexus between empire (variously defined), corporate oil and world oil status. Maugeri is an acute observer of the various connections involved.

Yet Maugeri's analytics are built less on a theory of empire than on the evidentiary proofs on which such a theory must be built. The lesson for such theory is that any model of empire must allow for dramatic shifts within long-term cycles. This comes into sharp relief whenever an old paradigm is breached and a new oil game-plan is established.

In considering the economics and related trajectories of oil empires, it is in the author's view instructive to inject into this debate some insights from non-linear dynamics and complexity theory. Such an approach requires deployment of the most advanced ideas in economics.[17] Here we should note that traditional neoclassical theory has well-known parametric limits. Pure market models have evident shortcomings. The global oil economy is much more complex than might commonly be imagined. The idea that it can be managed by any single entity is highly improbable. The economics of oil rarely exhibits neat linear progression. Oil cycles, responsive to world events and particularly volatile, are difficult to forecast. They are not easily calibrated over time. The struggle to understand such phenomena and to divine the secrets of the market has taxed many fine minds. Often the problem has been the underlying difficulty of modelling complexity and dynamics. It is akin to modelling the world, the unknown and the unknowable.

The economics of oil has been subject to evolutionary change – sometimes regular, at other times erratic, typically always complex. In this context we may say that the masters of pure economics have often built their theories on shaky ground. Though equilibrium may be useful as a concept, it is highly elusive in practice. Models grounded in marginal economics are often constructed on an assumption of *ceteris paribus* (other things being equal), yet in practice *ceteris* rarely remain *paribus* for very long. Steady-state models in general are ill-equipped to capture the dynamics of complex reality. In the oil game – as in chaos theory – both the macro (say, the empire of oil) and the micro (say, the corporate oil entities, including state players) matter. Each can have an impact on the other and act on the evolving paradigm. Bounded models (such as those found in Peak Oil theory) simply cannot correspond to world oil dynamics. We should resist their seductive influence even if they capture some of the public imagination. In this arena fashion proves a poor guide to understanding.

Models with predefined boundaries inevitably breed failure. In the real hydrocarbon world, exogenous shocks and uncontrolled variables abound. Our best approximation to the complex, dynamic oil world is probably the concept of "stable instability". The disconcerting truth is that there is no perfect, crystalline, theoretical cure for the diagnosis of imperfection. As a consequence, any model of empires of oil needs to be elastic and dynamic, allowing for maximal uncertainty. Nowhere is this more apposite than in the consideration of the (presumed) nexus between oil and war. The importance of this is that such relationships, complex as they remain, may be found to have shaped parts of the world order and at times the entire global energy paradigm.

✳

Numerous books have described the oil game in relation to world conflicts – past, present and even future. One notable recent example is Michael Klare's *Resource Wars*.[18] Klare's core thesis is that oil resource wars will be fought in future over our diminishing world oil reserves and supplies. It is a thesis built in part on flawed models of Peak Oil imminence. Here America's recent policies in relation to global security and military deployment are interpreted as pre-positioning for such a future. Hard military assets in Central Asia and the Middle East are regarded as the outcome of new strategies to protect access to world oil resources and crude oil flows to markets. The Russian démarche in similar theatres is seen as the flipside of the same coin. Klare also points to competing positions taken in the South China Sea by China, Japan and the ASEAN states. For Klare, it is not the mechanism of the market that will settle questions of scarcity and competing demands – it is military conflict.

For Klare, the priority given to oil resources results from a combination of factors. They include the dissolution of past ideological struggle, and role of oil in state wealth creation, econocentric security policies, escalating demand (especially from China), and both the presumed "reality" and expectation of oil scarcity. Though apparently persuasive, this case in fact rests on two questionable assumptions – that demand for oil is insatiable and that the Peak Oil thesis regarding oil supply is correct. What appears missing on the demand side is that price typically arbitrates demand/supply, while on the latter the end of oil syndrome has been greatly overstretched for its immediate and near-term or medium-term likelihood.

Although there are disputed oil zones, this has been true for some time. Even the ascription of exclusive economic zones (EEZ) offshore (a process in widespread negotiation now) will not necessarily cause military conflict, given that many overlapping areas across the world have been placed into joint development zones or similar arrangements. More could be created in EEZ areas.

Undoubtedly, though, some actual or potential reserves may be contested in future, especially since such conflicts often bring other forces into play – ethnic competition, for example. For Klare, however, it is oil that is the primary *casus belli*, both in the past and for the future. According to him, "Conflict over oil will erupt in the years ahead … as … almost a foregone conclusion."[19] American policy in regions such as the Persian Gulf, Caucasus, Caspian and even Africa is interpreted in this light. The epicentre of conflict is expected to be the Middle East, as the apex of a triangle of major strategic importance embracing the Caspian Sea in the west and the South China Sea in the east. America's strategy of attempting to diversify supply by developing countervailing regions, notably Africa, is expected to succeed only for so long.

There is no way to test such futuristic hypotheses, though few would dispute that the key regions identified by Klare hold some potential for conflict – especially in the Middle East. In the Caspian region, the politics involving oil pipelines is fractious and powers such as Russia and even China contest hegemony. Similarly, the exorbitant claims staked in the South China Sea are widely regarded as a possible source of conflict-to-come. There may be potential for confrontations in other regions too, including the equatorial zones around the Andean belt and Central Africa. But as in the past, the reasons for all these conflicts might be expected to be typically multivariable.

Klare, then, proposes a global geography determined by oil. Contemporary conflicts may be interpreted in this light. The long war in southern Sudan, for example, is in this light attributed to the struggle for oil. The problem here, however, is that in reality such conflicts have typically resulted from multiple and complex causes. It is often almost impossible to separate out one causal dimension and to ascribe each and every case to that cause. To attribute all to oil is a heroic assumption and largely ahistorical.

It follows that forecasting oil wars will remain a daunting task. We will treat such issues, and their contemporary manifestations, in the round in Chapter 4. Here we should note that while any theory of modern empire needs to include hydrocarbon dimensions and conflict, the difficulty of basing such a theory on Klare's particular thesis is that it rests on faulty Peak Oil assumptions.

A second notable contribution to recent debate is *America's Oil Wars* by a former CIA political analyst, Stephen Pelletiere.[20] Pelletiere seeks to explain the fundamental rationale behind America's overthrow of Saddam Hussein's regime. He argues, perhaps not so convincingly, that Saddam aimed to control OPEC and turn it into a fully fledged cartel. America's invasion was related to the desire to put a lock on oil reserves and supplies. Pelletiere's critique involves us in questions not only of ethics, but also of ineptitude. The law of unintended consequences is one that is often lost on policymakers and many an empire has slipped its moorings as a result.

The argument of a former spook like Pelletiere certainly should not be ignored. After all, intelligence agencies the world over monitor crucial issues influencing national security and economic prosperity. In this regard, the CIA is no different from the FSB (ex-KGB), MI6 or the national agencies of oil-rich states. The problem is that such agencies operate mostly outside the world oil industry. It is well known that, when they turn their attentions to oil, they have exhibited a tendency to get things wrong. Even corporate oil – with its own embedded networks of operations, insider history, intelligence data, high-level industry contacts, international scouts and experienced executives – makes mistakes. We need always to remember that, in an imperfect world, perfect knowledge does not exist.

Pelletiere concludes, on a theory which has wide currency, that America took advantage of 11 September 2001 to launch what was supposed to be a low-cost war in Iraq that would enable it to establish a military base there and to dominate, or at least influence, OPEC through its control of oil reserves. Saddam's own intention to dominate OPEC is seen as a measure that would have jeopardised American–Saudi management of the world oil system, regarded as a sort of usufruct over oil on the cheap. America supposedly feared a cartel dominated by Saddam's Iraq and an assertive Iran, both high-absorber states inclined to be price hawks. Its neoconservative strategy was, however, scuppered by poor planning, so that America is held to be likely to fail in its imperial quest, at least in Iraq. If indeed invasion of Iraq was just about the oil for the American government, it could not have made a worse job of its accomplishment.

In *Resurrecting Empire* Rashid Khalidi traces what his subtitle describes as "Western footprints and America's perilous path in the Middle East".[21] According to Khalidi, much of America's

quandary results from ignoring the history of Western intervention in this critical theatre of oil. America's own involvement
in Middle East oil has been long-dated. Middle Eastern history
reveals profound resistance to America's ventures (such as its
support for the Shah of Iran, for example). Any honeymoon
periods are generally short. The Cheney-driven neoconservative
lunge at Iraq is subject to the familiar problems. Although
America's ambitions for direct hegemony over oil may fail, some
military presence and indirect forms of influence over oil may
well long persist.

The Middle East is important not only for its oil reserves but
also for its strategic land bridges, oil choke-points, sea lanes and
pipeline networks. The history of the region is marked by asym-
metrical relationships between corporate oil and weak states. This,
though, is no longer the case: nation states are ascendant. The
Russo-British Great Game was played out in this environment
and competition between empires (monarchical or otherwise)
over oil has rarely been absent. The Seven Sisters lost control
to, as it were, a harem of new elites (autocracies, royal families
and nationalists) and their dominant state players (for example,
Aramco, NIOC, KPC and INOC). Over the past 30 years or so
these elites have gained control both of their states and, via their
national oil companies, over the regional oil game.

The raw power of America has not held undiluted sway in
this arena. Its difficulties portend the reshaping of an era and an
empire, though unbridled retreat is not to be expected. The Greater
Middle East remains a core zone both for corporate oil and for
those state oil companies with international upstream strategies

(the latter including Petrobras, Petronas, KNOC, Jogmec and the Chinese players). The Great Game has been transformed by the modern oil industry.

In *The New Great Game*, Lutz Kleveman details much of this contemporary history, from Stalin forward, with the trained eye of a journalist.[22] It is notable that so many empires, and former empires, of oil retain their own deep involvement in this region, along with corporate oil and the new state players. Kleveman argues that much of Central Asia, its patchwork of rivalries already installed, is still playing out the drama of Stalin's designs. The observation reminds us that the footprints of empires past can linger long into the future beyond their termination dates.

According to Kleveman, the "new Great Game over oilfields and pipelines in Central Asia gives but a foretaste of future energy wars over the world's remaining oil and gas resources".[23] This speculative view is shared by many, especially on the internet. The complexity of history, the power structures, alliances, clash of empires, competitive oil interests, the geography of pipelines, the presence of vassal states and minorities, rebels and oil mafia, and the influence of conflicts ancient and modern – all combine to make the Greater Middle East an unstable powder keg, and yet one critical to all competing world powers.

It is on Lord Curzon's Eurasian chessboard that America's search for dominance will confront the rivalry of, and possible "competitive collusion" between, Russia and China. They loom large in the world oil future. Their antagonisms, deep-seated but patched up in modern gas co-operation and oil deals, have set up a new pincer movement that may see vassal states (such as

Kazakhstan) and faraway empires (America) alike as temporary interlopers. The old "evil empire" is now the "energy empire" while the "middle kingdom" ruthlessly pursues its own agenda with scant regard to America or third-party interests. Both fret about strategic encirclement, the near abroad and the associated oil scenarios. Much dispute between these three superpowers in relation to energy focuses on Iran. Conservatives in Moscow and hardliners in Beijing are unsympathetic to American military-backed advances. In the circumstances, any American dream to reorder this world in a pure reflection of its image, or as a docile supply point for crude oil, may be bound to bite the desert dust.

Many recent works have focused on the question (explored in Chapter 4) of an oil/war nexus.[24] Some of these are catastrophist in nature and see oil as a unifactoral explanation of major geopolitical events as diverse as the collapse of the Soviet Union and America's so-called war on terror. In *A Century of War*, William Engdahl instructs us about the role of the Anglo-American empire in shaping the world order through wars. He advises that America's adventure in Iraq has had but one aim: "It was about oil".[25] The quest for oil is driven not merely by corporate greed, but also by a concern for geopolitical hegemony as a basis for national security. Engdahl locates American foreign policy initiatives squarely within the tradition of a unique oil-driven grand strategy, founded on historic theories about dominance in Eurasia as a key to control of the world.

These contemporary theories rest, however, on a fundamental flaw, namely the assumption that oil production has reached, or will soon reach, its peak. Peak Oil myths are often used to explain almost all American initiatives in the world. US foreign and military policy has aimed to control "every major and potential oil source and transport route on earth", so that America would be in a position to decide "who gets how much energy at what price".[26] America's strategic aim is apparently to gain the ability to deny oil to adversaries, such as China. Once military control is established, so such theories run, the energy dominoes would fall one by one across the world (*inter alia* Georgia, São Tomé & Príncipe, Libya, Sudan and Colombia). The American empire would then be able to influence, across the entire energy spectrum, its relationships with Japan, China, East Asia, India and Russia. The worldwide grip on oil flows would in this thesis be the real weapon of mass destruction.[27]

The fundamental lesson from all this is that, while a model based on empires of oil certainly helps us to understand the contemporary world, it is unwise to imagine – for imagination can go to extremes – that absolutely everything falls into a simple diabolical determinism in which oil-prone societies and oil companies lack any latitude.

Andy Stern's argument in *Who Won the Oil Wars?* over why governments wage war for oil rights is not too dissimilar.[28] It likewise is built on problematic Peak Oil thinking, considers almost everything to be about oil and treats war as an inexorable component of this dual drama. In this book (published, appropriately enough, by Conspiracy Books) the twin beasts of Big Oil and American empire are invoked, with the pursuit of oil tied to "greed, corruption and belligerence".[29] Together they stand accused of corrupting or subverting governments and of heinous crimes spanning the world's continents. China and India's new démarche into developing oil worlds is a rerun of the 19th-century scramble to colonise Africa. So, we are told, "We will see 'oil wars' increasing in frequency and violence."[30] The problem here is not that Stern does not succeed in highlighting many instances of oil's critical role in national strategies, but that the determinist outlook asks for all but oil to be put aside in understanding the past and in intuiting the future. Rigid linkages are drawn between empire and oil to exhibit a certain tautological determinism. It is as if to have empire we must have the oil, while to have and hold the oil is to have empire. Again we need to bear in mind the dangers of unifactoral determinism.

The portrait of the world oil industry in such works is at odds with the experience of those who have worked in the industry over the decades. It is also suggested that the global oil game is akin to "bribing the brigands", and even killing for oil, whether in Latin America, Africa, Asia or elsewhere. Industry insiders would not deny that operations have existed in contested conditions, or that some corruption exists, or that such practices may continue. Yet this certainly does not equate to the sum of industry strategies, behaviour and practice. In the modern era only a small minority of upstream oil deals might have been contaminated in this way. It has, quite simply, never been shown to be otherwise while it is notable that almost all such critics sit outside this industry looking in.

In the end Stern's views amount to a form of Oil Apocalypse Now, with much worse to come: "The oil wars of the 20th and early 21st centuries may be nothing compared to the battles of the future."[31] Once again, the foundation for such a belief is the Peak Oil mythology: our empires of oil will be caught up in violent struggle for diminishing oil resources; future oil wars may be averted only through dramatic cuts in oil demand and/or the mandated development of alternative/green energy. Neither is anticipated. Thus "the second half of the oil age will be bloodier and more violent than the first".[32] Unfortunately, while the depiction of the past is unifactoral, so the future is projected through a similar limited prism.

Crude Interventions by Gary Leech belongs to the same genre. A new world disorder is predicted.[33] The peoples of the South have their socio-economic and human rights trampled by the combination of corporate oil and militaristic American foreign policy. Once more, the forecast peak in oil production is seen as driving the race to control the world's reserves. Again, this presents us with a caricature of the world oil game.

According to Leech, the oil companies have sealed pacts with corrupt regimes (notably Nigeria and Angola) which contribute to their people's impoverishment. Colombia's new Independent Licensing Agency (ANH) has, Leech argues, provided contract conditions for foreign company concessions that lead to impoverishment of the people. He appears ignorant, however, of the reasons for Colombian upstream strategy which has been to resuscitate exploration and production. And in a simplification par excellence, Leech comes to the view that the South is "cursed with an abundance of the world's most sought-after resource".[34] In reality, oil is not a curse as such, although its mismanagement (typically in the hands of government) might, but need not, become one.

The common factor in studies within this genre is that the responsibility for the so-called curse of oil and any related state economic mismanagement is, in a move now resonant in public opinion, consistently laid at the door of corporate oil. Few take the trouble to acquaint themselves with the deep and complex histories that have shaped underdevelopment in the third world. Nor do many accounts, if any, consider on a cost–benefit basis the enormous impacts of corporate oil portfolio investment, without which many states (notably Nigeria) might be considerably worse off. Indeed this naivety among writers on oil, mostly from outside the industry, leads to much simplicity in explanation, error in contrast with the evidence, and flawed diagnosis.

Although some commentators exaggerate America's power, Leech is among those who believe that we are witnessing the decline of American dominance. He concludes his book with the view that the new global disorder typifies what happens whenever "a hegemonic power begins its decline".[35] This observation is worthy of note. There is indeed an emerging, anarchical fracture within contemporary globalisation; the American empire does seem, at least currently, to be in a phase of erosion; and conditions in several oil-producing countries appear decidedly blighted. But this is not enough to assume that the combination of America and corporate oil is either the sole or fundamental reason for global dislocation and poverty. It's as if nothing else matters. History teaches us a different vision of the past.

There are now numerous studies on the specific arenas in which it is argued that the American oil empire has been wilfully damaging in its impact. The Niger Delta is a particularly popular locus of study. The authors of *The Next Gulf*, for example, implicate the UK, America and corporate oil in this zone of long-term, low-intensity conflict, where conflagrations have played a role in the crude price upside over the past few years.[36] Active members of an environmental social justice group, their discourse focuses on the delta, the well-known Ogoni struggle against the Nigerian government and Shell, and the view that America and the UK are preparing a new century of plunder in Africa, especially in Nigeria. Paradoxically, the Chinese démarche into African oil and resources worldwide appears not to enter their vision. The travails of the delta communities are depicted and the armed gangs there are seen as a response to Western ultra-exploitation. Corporate oil and the Anglo-American empire are held uniquely responsible. The spin equates this triangle of interests with the old Atlantic slave trade. The contemporary oil trade forms a new version of the traditional triangular pattern of historical pillage. Investment comes in, oil goes out, and the delta is impoverished and brutalised.

By all the evidence the delta's oil history is not a pretty one. Successive corporate initiatives to ameliorate conditions with the band-aid of corporate social investment may even have made some conditions worse. In *The Next Gulf* Shell stands condemned for its alleged role in the execution of Ken Saro-Wiwa and, it is claimed, numerous instances of complicity with the military, police, state agents and assorted agents provocateurs. The company contests this interpretation. Nonetheless, now Shell's social licence to operate has been compromised, notably in Ogoniland where it has had to declare *force majeure* and later make an exit.

coalesce
nexus

Here the "war" with which oil is associated is not the classic military adventure but is seen rather as a protracted assault directed against the local communities in the interest of plunder. Companies are accused of a form of "clearwash", concealing their complicity from distant political publics (financial institutions, shareholders, NGOs, the media, customers, opinion-makers and governments) by means of sophisticated public relations and supposedly bene-volent social projects. In such contexts, corporate oil is, as we shall see in Part 3, often unsure which course to steer. The point is that for corporate oil the issues involved in producing oil in poor regions are often more pertinent to management than those raised by considerations related to classic imperial warfare.

The essential message presented, however, is that a New Gulf crisis is upon us, involving this time the Gulf of Guinea. All manner of American interests coalesce around this newly critical source of crude. A nexus between the military and oil is being formed. Indeed, the US might well be moving towards declaring the Gulf of Guinea an area of "vital interest". The establishment of Africom as a command in US military architecture in late 2006 signals heightened concern in that direction, but this appears some distance away from the dramatic as depicted in such works.

A number of other measures are seen as designed to secure regional dominance for the Anglo-American empires of oil. These include, for example, the UK–US Energy Dialogue on areas of shared concern in relation to energy, moves to wean Nigeria away from OPEC, and involvement in the Extractive Industries Trans-parency Initiative (EITI) governance and transparency process (often seen by many critics as a clearwashing arrangement). These initiatives are taken as "a means to an end, not an end in itself".[37] That is, they form part of revised imperial strategy. Clearly they could be read as consistent with Western interests.

Whatever angle is taken on these criticisms, it is evident that any adequate model of empires of oil needs to accommodate an understanding of oil within imperial strategies and of how corporate oil manages its material and non-material interests with respect to world and local operating environs. Many of the debates on these issues are at present framed largely by those critical of the interests of corporate oil and in a context lacking historiography and appreciation of shifting grand strategies.

In the literature there are also conspiracy theories attributed to the world of oil. These have spread widely, especially via the internet, and have come to command much public attention. Some derive ultimately from ideas found within the Peak Oil movement. At times, Big Oil is cast as some sort of *deus ex machina* wholly in control of the world oil game; at other times, OPEC is given the role of the Almighty.

According to some conspiratorial subplots, either a leading company, such as ExxonMobil, or some shadowy corporate collective is engaged in a vast conspiracy to deprive humanity of its rights and of the benefits that should properly flow wherever crude is found in abundance. Others take the US as the ultimate arbiter of the global oil game. One or other of these, sometimes both in tandem, are seen to orchestrate the money game that governs the distributive impacts of this rare and valued energy.

According to other narratives searching for villains, it is the Middle Eastern sheikhdoms or OPEC that are the source of all the problems in the oil world.[38] In *Over A Barrel* Raymond Learsey makes a sharp thrust at OPEC. He attacks its methods, its status (said to be in violation of WTO global rules) and the alleged manipulation of crude prices to the detriment of consumers (especially those developing economies bereft of oil resources). Rises in the oil price are taken as confirmation of OPEC's manipulative strategies.

Here we should note that the question of whether OPEC is in effect a cartel is a much-disputed point of economics. Economic theory would suggest that it lacks the key ingredients to be a true cartel: sufficient control over supply and discipline to enforce its rules. Overall, OPEC has enjoyed moments of grandeur, others of critical influence and some of powerlessness. Learsey, however, is unequivocal. He builds his case on the view that oil has never been scarce. Oil scarcity is said here to be merely a convenient myth, peddled by OPEC spin doctors and fellow travellers, and relayed by a litany of gullible analysts, consultants, uninformed pundits and dodgy politicians. Furthermore, a number of non-members, including Mexico, Russia and Norway, are accused of collusion. OPEC has managed its affairs in secrecy and, according to Learsey, some of its members (notably the Saudis) do not disclose their true oil reserves. Thus the result of OPEC applied strategy is a contrived shortage for which the world pays dearly.

As a putative price-fixer, OPEC is charged with breaching US law (though it is immune to prosecution). Learsey not only calls for an end to OPEC's policy, but also damns America for having struck a Faustian pact with this devil. In this caricature,

America has apparently sold its soul for a steady supply of cheap oil, at the expense of risks to its security because of its excessive dependence on OPEC. Some petrodollars, moreover, are believed to have wound up in the hands of stateless terrorists and leaders of rogue nations.

OPEC's agenda shifts with the tides. In *Over A Barrel* its central aims are said to consist of extracting maximal economic rent from the world's consumers, placating the G8 and dissuading corporate oil from investments in exploration elsewhere – a paradox since it is high crude prices that encourage such strategy. The security premium built into oil prices is attributed to hysteria and fear – psychological states manipulated by OPEC's skilful propaganda, mediated by traders of the electronic barrel. Corporate oil, along with the industry's media, is accused of self-serving collusion – which, despite the clamour for market stability, is seen as a one-way bet for OPEC (though in reality, we should note, changes in OPEC's surplus/deficit positions reveal that this has been untrue over the longer term).

Here it is enough to note that OPEC may act, or appear to act, as if it were a de facto empire of oil, except that its nature reflects a loose, sometimes fractious, interstate coalition. The interests of its members often differ (even in respect of the price of crude). Over time it has lost members (Ecuador, now returning to the fold, and Gabon), seen members weakened (notably Indonesia, no longer an exporter) and begun to recruit new members (Angola in 2007, perhaps Sudan in future as well). From late 2006 Venezuela sought to interest Ecuador in rejoining the club, an event confirmed in April 2007. In Learsey's view it is OPEC that has been the real victor in Iraq: the latter's quota has been unused, the fear factor has been driven higher, and corporate risk in world oil markets has been accentuated.

Paradoxically, OPEC's protector, at a cost of some $80 billion per year for security in the Gulf and for some members, is the very empire (that is, America) that it supposedly robs with finesse. Some have argued that if this cost were added to the barrel price, crude would be found not to have been cheap at all – still less now. For the time being, however, OPEC's existence seems not to be in question; but its calibrated capacity to manage world crude price levels has often been: for instance when it tried to apply the $22–28 per barrel price band a few years ago.

✳

The critics clearly disagree on which body is supposedly the ultimate arbiter in world oil. For some it is corporate oil, for some it is the American empire and for others it is OPEC. There are many devils from which to choose. Some explanations border on reductionism or mere fashion. Sometimes it appears that all our woes are the responsibility of Putin or Chavez and related allies. Others with less credibility even point to key oil executives. Yet it is most unlikely that any single source could ever account for the many configurations to be found in the oil world.

Many of the works in the popular press present a purely political or selective view of the oil game. Some have an evident political or ideological agenda, while corporate oil, state oil companies and governments have their own various interests too. Then there are competing grand strategy views on world order in which oil plays some critical role.[39] In some geopolitical studies, the role of oil is typically linked to the interests of diplomacy, statecraft and the strategies of warfare. In these works, however, oil is important but often made subsidiary to global politics, military strength and alliances struck between states in the balance of powers.

None of these traditional models – whether based on historiography, market forces, corporate strategy, conflict theory, conspiratorial models, or grand strategy design – provides a sufficiently broad or integrated lens to capture all the essentials and critical shapes found in and around the oil game. Each prism illuminates some aspects but obscures others. Partial analysis can yield only partial understanding. What then to do?

Our explanatory model needs to be broad enough to accommodate many critical features. These include changes in conditions over time, including paradigm shifts in the oil world. The framework should relate the state of oil to the state of the world – including the process of globalisation, the role of nation states, and the competing roles of corporate oil and national oil companies, with allowance for the impacts of the many antagonists aligned against the oil industry.

Without such an overarching model we will be left as prisoners of chaotic events. In this study, therefore, a holistic view is suggested. The aim is not to provide a "history of everything in oil", still less a deterministic history. Rather it is to provide a heuristic model of empires of oil that will allow space for dynamics, for cyclical and epochal change, statecraft, the origins and realities of competitive empires and their contrary claims, and the shifting tides of grand strategy in world affairs.

3

Empires and oil

Edward Gibbon's monumental study of the Roman Empire was published over 1776–88. It remains a classic study, though the interpretation of events has been much challenged and revised since. Gibbon provided a portrait of many aspects of ancient Rome, including the assault on the empire from within and without by an array of barbarian forces. These included the Vandals, Goths, Visigoths, Huns, Celts, Alans and the coalescing tribes drawn together from Germania.

The limitations of the long-enduring Roman Empire emerged in numerous, sometimes subtle, ways. They included military constraints, problems in co-opting the barbarians, periodic rebellions, divide-and-rule strategies with faulted alliances, the split of empire into western and eastern hemispheres, the emergence of threats from fundamentalists (here Christians), the difficulty of managing diffused or only partially integrated non-Roman cultures and the eventual arrival of barbarians at the gates of Rome. There were difficulties in sustaining the imperial heartland as regional stretch diminished capacities in critical areas. Political intrigues and diplomatic issues undermined Rome's grand strategy design.

Many exogenous shocks were induced by barbarian counter-strategies that proved to be both sophisticated in effect and durable. For all the empire's evident strengths, its construction, nature and management ultimately generated fatal strategic weaknesses.

The analogies between the Roman Empire and modern empires of oil may be seen as profound. States that are reliant on oil, especially foreign oil, must ideally have at their disposal the durable legions of corporate or state oil. Reliance, on the part of state oil companies and to a lesser degree by corporate oil, on state initiatives provides telling parallels with Rome's legions in the outer world. We can draw on such analogies in constructing a model that integrates diverse and at times conflicting micro (for example, oil companies) and macro (for example, empires of oil) phenomena.

The way in which empires have been eroded and displaced by competing empires provides a useful template for the periodicity of resource nationalism allied to state–corporate oil interests. A company's decisions on hydrocarbons investment, upstream access and portfolio strategy are often mediated by aspects of the player's state of origin, especially its geopolitical dispositions and ultimate reach. Here a large-scale model founded on the notion of empire appears broad enough to incorporate the many historical, partial, chaotic and temporal models used to explain the world's upstream industry and related shifting dynamics.

Myriad instances from oil historiography demonstrate the virtue of taking empire as the paradigm. As we have seen, the industry was built on Rockefeller's personal oil empire. The Seven Sisters reflected an apogee of Anglo-American and European dominance. The emergence of the industry and its connections to the British empire, later displaced by America's oil ascendance, have been well demonstrated in the analytical and historical works of Odell, Yergin, Maugeri and others. Today America commands an empire of oil that still probably dominates the Western world, even if it is challenged on many fronts both on the outside and by some corporate oil entities found inside the West at large, notably companies from Europe.

The rise and tribulations of OPEC reflect a global counter-strategy that reflects the world that its members first encountered. Oil's history has been closely linked to conflicts between rival empires (British, American, Russian, Ottoman, German, Japanese and now, among others, China). Supplies of hydrocarbons, especially crude oil, have been a *sine qua non* for geopolitical dominance in modern times, especially since the Second World War. In the struggle for command over supply, several states (perhaps, most notably, Russia, China, India, Iran and Venezuela) have begun to emulate many aspects of the behaviour of older empires.

The hallmarks of the contemporary oil game are national strategies of ownership, resource access/denial, petro-diplomacy, preference in deals, energy bilateralism and oil security initiatives. A major manifestation today is the challenge to corporate oil from national oil companies, with intensified competition found within a much longer corporate oil tail. The faces of the politicians change more quickly than the processes with which they are involved. The various empires confront many common problems. Each empire of oil encounters constraints from the existence of others and from the amorphous power of the world oil and capital markets. No single power easily predominates in this battle for ascendancy. Each faces a clash of interests and strategies overlaying competition for power between diverse political formations and civilisations in rapid transition.

Let us consider in more detail the analogy between the modern oil world and the Roman Empire. We will concentrate particularly on the challenge from the barbarians – bearing in mind that barbarians (seen here as threats to the old order), notably modern variants, may sometimes be encountered within the empire.

Peter Heather's *The Fall of the Roman Empire* helps to remind us of the way in which empires may be subject to exogenous shocks.[40] As explained, "The Roman Empire was the largest state western Eurasia has ever known. For over four hundred years it stretched from Hadrian's Wall to the River Euphrates." The victory of Marcus Aurelius over threatening Germanic tribes was an epochal moment in cementing Rome's hegemony but, as Heather writes, "two hundred years later, the Romans were still at it". In AD 357, 12,000 of the emperor Julian's Romans routed an army of 30,000 Alamanni at the battle of Strasbourg. Yet within a generation, the Roman order was shaken to its core and Roman armies "vanished like the shadows". Gothic refugees were granted asylum by Emperor Valens. Within two years the barbarians allowed within Rome's borders revolted, defeated the Roman army and killed Valens. The empire never recovered. In AD 476 the last Roman emperor in the West was deposed. The descendants of the Gothic refugees formed the military core of one of the successor states, the Visigoth kingdom.

According to Heather, Rome had not been on the brink of moral, social, or economic collapse as Gibbon and many since have assumed. It collapsed because it underestimated the barbarian onslaught. Its military prowess and its ideology of superiority created a conquest state, but it was one that became bureaucratic, slow-moving, prone to the unequal distribution of internal benefits, and subject to regular tensions and war. It was to prove all too vulnerable to the barbarians.

Today, the American empire dominates the globe on many but not all criteria. Yet it too faces problems. Shifts following the end of the Cold War together with the rise of competing states threaten to erode American power. As Kagan has argued, a de-linking from Europe in some strategic interest and behaviour is now evident. As with Rome and Constantinople, the interests of the two blocs may not be the same. Certainly their threat matrices concerning oil and energy have diverged. Perhaps the West is simply ceasing to exist as a coherent entity, or requires reinventing.[41] America seems to have underestimated the enormous challenge of its Middle East projects (most especially that of producing a democratic order modelled on its own precepts). Its lack of a classic imperial past, at least one with directly managed worldwide colonies, may leave it ill-fitted to establish a Pax Americana. Just as America's vulnerability over oil has come to the fore, it has been caught out by the rise of alternative empires of oil in various forms around the world.

The longevity of Pax Romana was created from a patchwork of moral standards, cultures, new religions (especially Christianity) with the overlays of shifting boundaries and unstable hierarchies. It was barbarian infusion into the empire and invasion from without that was to prove crucial in the collapse of the empire. The invaders, pushed by ascendant Hunnic power and migrations from Germania, pressed diverse barbarians into Roman territory. The collective scale of the various barbarian armies was far from insignificant. In the long march towards its end, the empire slid rather than toppled. Instability at the imperial centre allowed the barbarians to exploit diverse strategic opportunities to advantage.

As Gibbon has shown, significant limitations on Rome's military, economic and political capacities were a condition for ultimate collapse. Streams of barbarians carved out niches within and around the Roman body politic. The power and wealth of the Romans became a target for external assailants who deployed Roman weaponry to sound effect as adjuncts to home-grown military skills. The scale of empire generated an asymmetry that in time led to the erosion of dominance.

America's empire of oil has passed its zenith and must now continue to acclimatise to a reshaped world environment. It must deal with a mixture of oil regimes with stronger local and regional influences. The erosion of the empire is not primarily the result of internal factors. It has more to do with barbarian ascendance – from Russia to Venezuela to fragile and autochthonous regimes across Africa, Asia and even in Latin America. Some now take a non-Western or even anti-American perspective. As the story of resource nationalism reveals, the rise of each new oil domain encourages others to create or grow mini-empires in oil.

As the oil world has widened, so the space permitted to

American companies has shrunk. Self-induced sanctions – for whatever moral or political reason – have not been to the advantage of American oil companies. The collective power of OPEC has grown, and a host of threatening states now command the attention of US diplomacy and military forces – partly connected to the question of access to oil. Corporate oil – originally an Anglo-American edifice, now broader in scope as an OECD/EU venture – faces multiple competitors, as powerful players (including those with state support) emerge in the barbarian energy world, many targeting the same oil potential sought by Western companies. The barbarian worlds within and around global oil are as multi-faceted and complex as those of Roman times. So our model must allow for the new intruders that have insinuated themselves into the world oil order.

In what is a masterly revisionist study, *Barbarians: Secrets of the Dark Ages*, Richard Rudgley has shown how barbarian forces shaped the Roman Empire and, especially, its demise.[42] In effect, some inside became integral to its survival and later its wilting. We might ask how such an interpretation can help us understand the oil world of today. It has been a thesis of many colonialist writers in settler societies that the masters of the house stood apart and beyond, superior in status to the servants (that is, the barbarians).[43] But masters often depend on the servants. As in Roman times, the aristocracy's views on lesser societies fed off and constituted a one-dimensional caricature. Inferiors were seen as primitive, ignorant and dangerously unpredictable. The colonial world has likewise passed by, although rather than completely collapse it has transformed itself into myriad polities.[44] Thus did Rome cede hegemonic power to new rulers, states and societies which endured as such for many centuries.

As Rudgley shows, the barbarians competed for power among themselves, while also challenging the superpower of the times. They formed no homogeneous entity. Their identities and strategies were diverse. Some lived within the empire, others beyond its borders. Their own narratives did not conform to those of the Roman scribes or spin doctors. The barbarians understood the power of empire but continued to regard their own cultures as central. The *mission civilisatrice* typical of colonial powers, aimed at encouraging a mixture of assimilation and subjugation, failed to change the fundamental, long-established social realities. This is not unlike our modern world.

While (at least until Constantinople) all roads led to Rome, the barbarians always sought to assert their own rights and interests, even though the lines between masters and servants would sometimes blur. A plethora of barbarian tribes grew like weeds amid the fields of the Roman world. The strategy of Hadrian to divide the Romans from the barbarians by building a wall at an outer fringe of empire ultimately failed. The margins of empire were always volatile and porous, beyond the direct control of the Roman senate.

In the contemporary oil world, many states now contest the world's oil patrimony, and national oil companies compete against corporate oil. In the process they refine their shape and portfolio, each forming its own strategy. Their apparent bewildering array of strategies sometimes foxes corporate oil and those politicians who seek to maintain the old order. Many oil executives that the author has met imperfectly understand this complex world inhabited by the "corporate other". The barbarian players, however, know their strengths and weaknesses, seeking either to collude or to dispense with corporate oil as opportunity offers, a lesson well appreciated in the Kremlin.

By its end, the Roman Empire became rule-bound and insufficiently flexible for its own strategic necessities. It was unable to control its essential source of sustenance: the granaries of North Africa. Roman mythology failed to integrate and manage the mosaic of nomadic opponents, including the Goths and the Huns who rejected ultimate integration with Roman culture, and the Roman leadership continued to operate with distinctive mindsets that misconstrued barbarian strategy. The empire needed to juggle many ethnic groups to maintain control. At times its forces were required to prevail in too many places at once. In the end, the Mediterranean stranglehold on the ancient world succumbed to assault from a barbarian hydra.

The primacy of the world's current superpower has been, along with its Europe counterpart, rule-driven and ethics-prone. Its ethos finds some modest correspondence inside many developing worlds. Its proclamations for universality are resisted in many quarters. Some corporate oil players are constrained by sanctions, ethical issues and shareholder concerns. This may further limit their scope to operate in future barbarian oil worlds. The issue at the top of America's policy agenda is the nexus of oil security, resource access and continuity in the supply of affordable crude – the *sine qua non* of any modern empire. The West has certainly not succeeded in dominating the new world oil game across all continents. Indeed, even its position as first among equals in the hydrocarbon world is in question.

It would be wise then to be sensitive to the forces that shape the form and destinies of empires. Here we may draw on the historiography of Niall Ferguson, especially in *Colossus*, a critical and seminal study of the rise and fall of the American empire.[45] Ferguson is in no doubt that, whatever American commentators themselves may say, there is indeed an American empire. Although he does not focus on the oil world, the analysis has much significance for us here.

America's predicament has much in common with that of the later Caesars, its empire holding more than military dominance. It embraces within its ethos a universal civilising mission, as well as a strategy for the diffusion of freedom according to its own definition. Only the most naive of America's opponents could be surprised by the nexus of military profile and hydrocarbon interests. The US has long targeted as crucial access to the vast reserves of the Middle East. Although there has been much debate over the role of oil in the invasion of Iraq, Machiavelli would surely have discounted such debate as so much hot air. Iraq has the world's largest undeveloped proven oil reserves: it would have been out of character for America, as an empire of oil, not to have noticed, but its actions may not have been consistent with this interest alone.

From America's perspective, the establishment of world order entails what may be called "state replications". This image of the world requires states which, if not actually ruled by America, at least govern themselves in the "American way". This vision requires an opening of world markets, including those for oil and energy, even if in some domains the US is often protectionist. With the exception of oil sanctions, any closing off of oil resources, through ultra-nationalism, lockouts and so on – oil sanctions excepting, of course – simply does not correspond to America's view of the best interests of itself, its oil companies, or its allies.

In world history, Ferguson shows that apolarity, or the existence of a singular world power, is an aberration. Typically, rivals emerge. Empires usually nurture within themselves the seeds of their eventual decline. The early 1990s may have been a moment of such revelation as Cold War victory turned towards a new de facto multipolarity. Since then, the development of fractured globalisation has made apolarity look anachronistic. We are now in a contested world no longer defined by the bipolarity that marked the Soviet–American struggle for hegemony. It is in this context, and largely within the 21st century, that newly ascendant national oil companies have exploded onto the oil scene in a resurgence of resource nationalism which has gripped even smaller states such as Chad, Bolivia, Ecuador and Venezuela.

The incipient erosion of American empire, in Ferguson's view, entails diverse trajectories leading towards global disintegration, limited supra-nationalism in world institutions and the rise of non-state actors in world affairs. In this context, corporate oil has had to pay a heavy price. It is faced with new oil contract regimes, much higher taxes, mandated decarbonisation, renationalisation, higher costs of compliance with new social norms and legal systems that at times threaten its material asset base.

Corporate oil may even need to design and manage compacts with UN agencies and (as shown in Part 3) almost everywhere to engage one way or another with a plethora of NGOs (some benign, some hostile).[46]

Conflicts between empires (some 70 or so spanning world history) have much centred on finite resources, of which oil has now become the most important. The Russian empire that succeeded the Soviet equivalent is by no means at ease with itself. The conflict in Chechnya is more than a blemish: it is significant for its position astride important Caucasian oil routes. Islamic ascendance in various forms in the Middle East and Iran creates heavy weather for Western players in their search for oil access in a region where exploitation opportunities for corporate oil are far from untrammelled.

The US, though at a conscious level a republic, is according to Ferguson unconsciously an empire, somewhat inept in its imperial destiny and potentially likely to perish. It is seen by many as a Goliath in denial. As its claim to economic centrality weakens, oil supply emerges as a candidate to become the slingshot that might damage it.

The American way entails reliance on corporate engagements, secular bodies and local elites, with only a limited appetite for territorial expansion. Occupation or involvement in Iraq may prove a temporary exception to this generalisation. For all Iraq's upstream potential, at the time of writing no US oil company had taken a significant position there. Corporate oil is called upon to go forth and multiply its assets worldwide without US central planning to guide its strategy. Those who argue that Big Oil is hand-in-glove, always and inexorably, with the American government miss this notable disjunction. Corporate and government agendas are often liable to diverge, and have done so in the past and present. It is therefore critical not to rely on some deterministic model of empires of oil in which the interests of the former simply dictate the policies of the latter.

Even so, the US harbours hundreds of military installations strung across many countries at this time. Its global military architecture spans the world and space. Its commanders can be compared to imperial proconsuls. Yet its power is far from unbridled. Recent apolarity followed by challenges on several continents has perhaps led to some critical resources becoming overstretched. The burdens of empire in a world replete with failed and rogue states are great. In this context, military–oil relationships across the world should not be seen as aberrant. It is a fact equally reflected in Russian dispositions in Central Asia as well as China's targeting of Central Asia and the South China Sea, and its search for worldwide equity oil and gas.

Though on a smaller scale, Nigeria's military and political endeavours in the Niger Delta may be viewed in the same light. They emanate from the protectorate of the Nigerian government's oil world and its intent to command this essential oil sphere. Regardless of moral claims or positions on human rights, such strategic actions flow organically from the natural order of imperial or state security interests in oil patrimony. Oil has become the *fons et origo* of survival and success for such petro-reliant states. The umbilical relationships between oil, state and military will consequently persist even while their forms alter over time.

Nowhere is this more true than in the ruthless, ultra-competitive, upstream world. As anti-globalisation sentiment has grown, American and European corporate oil players have increasingly become targets in the eyes of hostile organisations. Despite America's enormous cultural influence in an era fascinated with affluence, the country's "soft" power has much eroded. As a consequence, we should incorporate into any model of world oil an explicit role for the organised antagonists and enemies of the hydrocarbons industry, with corporate oil as its most visible target. Increasingly this will affect large and small oil players, including in time the national oil companies.

Because of its diminished power, many companies have sought to build their security independently from the formal empire of oil, taking risks to achieve equity ownership of reserves. Some independents have entered difficult domains on their own with no support from their home state and even, at times, against the declared interest of their governments. In looking beyond the state, companies with foreign portfolio strategies have sometimes in effect created foreign policy through their positioning and deeds – sometimes in compliance with widely accepted operating principles concerning transparency and governance and on occasions not (thereby attracting the unwanted attention of belligerent NGOs).

US dependence on Middle Eastern oil is now viewed as a strategic deficiency, a malignant condition that has grown apace in the post-Soviet epoch. The demise of the Soviet empire changed the oil world's cartography irrevocably. It led to the opening of Central Asia. Some hot wars in Africa duly ended (notably in

Angola) as ideology fled in the face of naked commercial interest. The later privatised Russian oil companies such as Lukoil became both competitors and allies to Western corporate oil. ConocoPhillips (COP) is now a principal holder of equity in this giant Russian oil conglomerate; BP entered Russia via large-scale acquisition in TNK-BP and with field deals; and other super-majors such as Shell and Total made investments to take PSA oil and gas concessions. All this was inconceivable only a short while before. Much of this pick-up in then-frontier portfolio has now been contested by a resurgent Russian state. More might still succumb to Russian strategy.

Meanwhile, however, in the Middle East the threats to America's role have deepened. An uneasy nexus now marks its traditional relationship with Saudi Arabia, once the bedrock of certainty. Even so, Ferguson sees "the threat to America's empire ... not from embryonic empires to the west or to the east ... but from the vacuum of power – the absence of a will to power – within".[47] It is unlikely that competitive worlds can be discounted. For these and exogenous reasons a diffusion of world power seems most likely. The resolve of the masters may be at issue and the consent of the servants has certainly dissipated. In the paradigm that is emerging, some role reversal has already taken place.

As Ferguson has perceptively argued, empires have tended to drive history, yet their lifespan has tended to decline, most dramatically in the 20th century. Rome's Western Empire lasted 422 years, its Byzantium successor 829 years. The Ottoman and Portuguese empires each clocked up approximately half a millennium. Others did not survive so long. The British, Dutch, French and Spanish empires endured for approximately 300 years, the Soviet empire just seven decades (a time-span not yet equalled by China) and Hitler's Reich (a military venture mainly) had a lifespan of only 12 years. Some ghosts of empires past continue to stalk the earth, with the remnants of lost imperial designs now evident in the histories of various nation states and ex-colonies.[48] Indeed, the boundaries for most states still date from imperial pasts. The scramble for Africa (after 1884) established the state imprints still critical for most onshore and offshore delineation for acreage on that continent, although many contested zones now litter Africa (in Somalia, Democratic Republic of Congo, Côte d'Ivoire and elsewhere) as late 20th-century balkanisation has continued apace.

Many empires were the unwitting architects of their own downfall. Territorial ambitions attract rivals and dissent. America's de facto empire, with its roots in the 19th century, may be prone to the usual causes of dissolution that include military deficiency, financial costs, public disdain and lack of will. These dynamics play out equally in the great oil game.

In terms of its oil requirements, the American empire needs ideally to procure supplies at costs below best options in an open world market and/or from competing oil empires. Security-driven oil procurement is not always the best strategy. The liabilities incurred through global military involvement are large. They are a factor in any assessment of the costs or benefits of relinquishing supremacy. So far as crude is concerned, America is less in charge of the relevant dimensions than might be assumed. Oil in the ground does not everywhere lie in its ambit of control or direct access.

So it has begun to consider strategic options, at probably higher long-term cost: biofuels, ethanol, shales, oil sands and so on.

The decline of modern empires has been closely associated with the instability of our fragile times. The ensuing shifts in the tectonic plate of geopolitics have brought with them ethnic disintegration, economic volatility and contestation. Self-determination has often brought aggravated conflict. This has been true especially of post-decolonisation African history, including that of such oil-driven states as Nigeria, Chad, Sudan and Angola.

The phenomenon of failed states too is a related and direct consequence of the abandonment of empire. Sometimes this has been entwined with what is called the "curse of oil" – though most often the curse stems from state economic policies, mismanagement and poor governance rather than from the geological endowment itself. The maps in the conventional world atlas are marked with lines neatly demarcating nation states. The internal and external conflicts associated with failed states mean that the pattern on the ground is nowhere near so visible or tidy. Yet the task of rebuilding such states is one that few wish to bear. The success rate, after all, is low – though some states, notably Mozambique, have chalked up some success.

Ferguson, like many, sees the Middle East as the focal point of future conflict, disintegration and a possible reordering of the grip of American empire.[49] Clashes between Islam and the West (*à la* Samuel Huntington) are viewed by some to overlay the patterns of imperial competition in the region.[50] It is an arena that attracts a resurgent Iran in its quest for regional dominance. Local contestation appears from the oil-rich Sunni Arab states with large hydrocarbon reserves (principally Saudi Arabia). It also produces clashes within many oil-rich societies, especially where (as with Kurdistan) the presence of oil mixes with unresolved ethnic disputes and unsettled territorial claims.

Other grand strategists see Eurasia, a vast zone that includes the Middle East and Central Asia, as the critical chessboard on which global powers inevitably compete. Zbigniew Brzezinski has argued that it is this delicate, oil-rich, arena that American power needs to manage in order to prevent the emergence of any new, large-scale, antagonistic world power.[51] From this point of view, recent trends do not offer much ground for optimism about any easy path to dominance. In Central Asia, Russia is resurgent. Here China is ambitious and its sphere of influence is growing. The Eurasian region is also an intense corporate oil battleground, in which state companies are becoming more powerful. American hegemony in the region may be described as shallow.

Although Brzezinski's view is instructive, it by default discounts much of the rest of the world. This is especially true of Africa, where a subordinate role is presumed. In fact, Africa matters a great deal to the US and is likely to matter more in the future. By 2025 Africa is likely to provide 30% of America's imports of crude oil. In December 2006 the American government signalled its intent to create Africom (the US Africa Command) as a means to rationalise its trifurcated European, Asian and Middle East command structures.[52] US policymakers now regularly include Africa on their radar screens, and it is much to do with oil that has attracted their renewed interest in a continent long neglected and still weakly understood in the US.

Overall, however, America has not been able to prevent collusion among rivals, maintain security and dependence among all vassals, keep tributaries pliant and protected, or prevent the barbarians from coalescing. This is precisely what has happened in the global oil game. There is, for example, greater collusion and partnership between national oil companies than ever before. Interstate hydrocarbons deals have proliferated. Oil supply strategies have diversified. The upstream industry is becoming less dependent on America. Old imperial allegiances have less importance in determining alignments within the overseas oil world. All over the world, access to barbarian oil is at a premium as bonus payments and bids attest. Many states look increasingly to each other to develop their energy futures. The ultimate lesson of our age may be that America may have been, as Brzezinski reminds, "the first, only, and last truly global superpower".[53]

It will take a Herculean global effort and propitious circumstances for a unipolar America to retain future world dominance, at least in oil. Contrary to Francis Fukuyama's prophecy in *The End of History and the Last Man*, the passing of the Cold War has not brought about a peaceful proliferation of parliamentary democracy.[54] Deep history just does not magically disappear. Indeed, models of empire have become more relevant to an understanding of our future. It is empires of oil that have been proliferating.

Earlier in Chapter 2 the argument was advanced that a model of empires of oil must be multidisciplinary, flexible and non-deterministic. Here we may add that it needs to find room for the complex roles of diverse barbarians in the modern oil world.

We have noted in particular the way in which barbarians may be significant for their actions within empires as well as from outside. Later Part 2 maps the myriad terrains of the barbarians while Part 3 develops a typology of modern barbarian threats inside and around various oil milieux. Before considering these dimensions, however, we need to investigate in more depth a number of themes interwoven around the relationships between oil, social conflict and wars connected to hydrocarbon environments.

4

The "oil curse" and social conflict

Oil is rarely seen as an unmixed blessing even though its endowment is a gift from the gods. It can provide great wealth for investors and drives the modern economies that enable higher standards of living. Yet oil is often associated with poverty, inequality, conflict and war. Many areas of the world from which oil is produced – Africa but one, Central Asia another – are among the poorest and most troubled on the planet. These associations have led many commentators to assume that oil is not a blessing, but a curse. Indeed, Nicholas Shaxson, a journalist, has claimed that it is oil itself – the corrupting, poisonous substance per se – that is to blame for most of Africa's ills in oil-producer countries. So Shaxson notes, "ExxonMobil likes to say that there is no resource curse, just a government curse … They are wrong: the heart of the matter is not rulers' corruption or companies misbe-haviour but oil and gas itself."[55] Reality is undoubtedly much more complex than the much-used and abused "oil curse" slogan suggests. Yet it is a chronic failure of analysis (in theory, history, economics and social context) to ascribe such ills as are found to the existence of oil. So it is highly relevant to investigate oil's relationship to war, conflict, power and economic hardship (especially macro-policies) – all elements that shape empire – especially with regard to questions of their incidence, inevitability, causation and implications for the world oil industry now and in the future.

✳

The end of the Cold War brought much talk of a "peace dividend" among the "chattering classes" and the Western intelligentsia. It was short-lived, if even real. Despite a multitude of initiatives designed to install and promote peace – international peace-keeping forces, security watchdogs, NGO campaigns, risk identification and early conflict warning systems – the course of the 20th century remained violent and bloody to its end.

Much empirical analysis of war has been conducted. Data sets have been collected and conflicts calibrated (by, for example, the Stockholm International Peace Research Institute). Deaths and injuries have been enumerated, costs calculated, massacres measured, genocides investigated, causes evaluated, the emergence of conflict modelled and simulated, behavioural variables correlated, crisis dynamics assessed, government failures identified and protocols proposed – all, it seems, to little or no avail in practice.

The last decade of the century witnessed over 50 major wars – two dozen or so of them active at any one time. Recent history has of course seen some peace settlements – including Aceh, Angola, Bosnia, Democratic Republic of Congo, East Timor, Liberia, Sierra Leone and Sudan – yet new wars or armed conflicts have arisen elsewhere in their place, and some zones of partial settlement have receded into renewed conflict. Many have had little or no association with oil.

Yet there has long been a presumed association between oil and conflict. According to some commentators, there have been four major oil wars: the Suez conflict of 1956; the Arab–Israeli conflict of 1973; the Gulf war of 1991; and the war on terror, including the invasion of Iraq. There have of course been major conflicts in which oil has been a critical factor. Perhaps the most brutal of all recent wars was the Iraq–Iran war of the 1980s, in which oil was present if only tangentially as the antagonists managed to find numerous rationales for their belligerence.

In view of this history, many commentators argue that there is an unavoidable nexus between oil and war. Moreover, it has been implied that oil has been closely associated with many other types of conflict besides classic warfare – with massacres, managed famines, guerrilla struggles, pogroms and genocides. Even when war recedes, bitter conflict over oil – its institutions, governance, patronage and so on – often persists. It has been argued that, although oil has enabled modern societies to develop, it has also, through its association with conflict, proved something of a curse.

Here we examine the relationship between oil and war first by surveying recent and contemporary history, especially in the developing world, and then by considering more general and generic issues related to these continents. It will be found that no simple causal nexus ties oil to wars and hence inevitable conflict: many dimensions have marked most struggles even within petroliferous states.

✳

Africa has experienced many types of conflict in recent history. There have been numerous *casus belli*, including religion, state autonomy, ethnicity and local grievances, but oil has certainly been a factor in some conflicts (for example, Sudan). There have been both intrastate and interstate wars, some with direct foreign involvement and some without. Consider first the major intrastate conflicts in Angola, Algeria, Sudan and Congo-Brazzaville – all oil-producing states – and some of their flawed transitions connected to the shifting profiles of competitive empires.

The 27-year conflict in Angola since independence in 1975 resulted from many factors, many unrelated to oil stakes directly: including Cold War struggle, internecine power struggles, ethnicity, regional power issues inside southern Africa and ideological strife. Oil, together with the diamond trade, played a key role in financing the conflict as it evolved, and has helped shape the character of the Angolan state that has emerged. The conflict affected the exploitation of onshore fields and exploration there. The need in particular to clear landmines continues to delay operations and hinder the opening of interior basins. In the enclave of Cabinda, where social conflict has continued, there have been negative impacts in the past on Chevron (there formerly as Gulf Oil) and regarding security for new entrant independents such as Roc Oil in the onshore. The idea that the Angolan war was a pure oil war does not stack up with historical evidence and the complexities of southern Africa at the time as related to South African government strategy and the role played by insurgents such as UNITA (contesting the MPLA) and

SWAPO of Namibia with sanctuaries then in southern Angola. In large measure conflict in Angola arose from the abandonment of empire by Portugal, fuelled further by competing world powers: America and the West against the Soviet Union, with apartheid South Africa's then-hegemonic ambitions also in play.

Since 1991 Algeria's government has contested threats to its authority from groups such as FIS (Front Islamique de Salvation) and the GIS (Groupe Islamique Armée), as well as social protests from the Berber community. Core issues included autonomy, autochthonous rights, military government, state control and inequality, and the struggle between secularism and religion. The presence of vast oil and gas resources may not have escaped the notice of the protagonists, but the ideological drivers for the rebel efforts were not built around the oil game itself. The geographical location of the reserves and areas of greatest potential in the distant Berkine Basin and south-western desert (as well as, now, offshore) did much to limit the negative impact on the oil industry. For most of this conflict, Western oil interests were protected by the Algerian state.

In Sudan, there was conflict in the south for over 25 years before America brokered a "settlement" between the government and the Sudanese People's Liberation Army (SPLA). The region affected included the former Chevron Heglig and Unity oilfields. This was a highly complex struggle involving multiple issues of autonomy, complex ethnicities (Nuer, Dinka), Arab–African racial issues, rivalries for political power and north–south religious conflict over *sharia* law, including differences between Muslim communities and the Christian/animist south.

The investments of several oil players were at risk during the conflict. Sanctions were applied to Sudan. Talisman Energy was pressurised to sell its equity position in GNPOC, Lundin sold some assets and OMV decided to withdraw. Other players ultimately benefited from the sanctions applied to Sudan. They included Asian national oil companies (notably Petronas, CNPC and ONGC Videsh) and a slew of private players from the Greater Middle East, including Zaver from Pakistan, Al-Thani from Dubai and even later PetroSA. The Comprehensive Peace Agreement to split oil revenues 50–50 between north and south has set the stage for a wider set of competitors to enter.[56] Some areas remain contested, notably the acreage claimed by White Nile (listed in London on the AIM) and long held under *force majeure* by Total, Marathon, Kufpec and Sudapet. The conflict and quasi-genocide in Darfur, itself highly complex in origin and character, continues to prevent the complete de-sanctioning of Sudan and so blocks the entry of many corporate oil players from the Anglo-American and Western worlds. Again the causes, drivers and core elements in these conflicts have not been purely of oil origin. Much at issue has related to a mini-empire in oil sought by the Sudanese government.

In Congo-Brazzaville during the 1990s there were serial conflicts between various ethnic groups and political factions. They escalated again at the end of 2003. The control of oil, though not the only consideration, was seen as the major prize. Yet a rendition of this history would show that many political, regional and ethnic issues fuelled these fires. The conflict thwarted the efforts of some companies then in place, and as a result during this time Congo paid a price in fulfilling its oil potential. The remnants of Francophone hegemony in Africa, connected to competing elites and ethnicities, played a part in this drama.

In the post-Siad Barre era since 1992 Somalia has fragmented. Conflict between clans and the transitional federal government (TFG) in Mogadishu has been unremitting, continuing in 2007 with the Union of Islamic Courts (UIC). New statelets – Somaliland, Puntland and at times in the area around Baidoa in the south-west – have long been formed. Oil assets both onshore and offshore have been frozen by *force majeure*. Renewed conflict in 2006–07 between the UIC (supported by some clans) and the TFG (aided by Ethiopia's military) had several ingredients, including many unrelated to oil. They included ethnic rivalries, armed militia, *sharia* law, jihadism, US support for Ethiopia as proxy in the war on terror, Ogaden rebel involvement and divisions between clans. The Ogaden National Liberation Movement even attacked gas fields owned by Petronas in April 2007 inside Ethiopia, killing 65 workers, mostly Chinese. The conflicts have clearly had a negative impact on the exploitation of Somalia's oil potential. A few small independents have taken acreage in Somaliland (for example, Ophir Energy, partly owned by South Africa's Mvelaphanda), with a few smaller players such as Range (with farming partner Canmex) present in Puntland. The failures apparent in this failed state have had minimal hydrocarbon connections, its balkanisation likewise. Indeed, the failed state of Somalia followed the demise of US–Soviet global antagonism and what became seen then as the strategic irrelevance of the Horn of Africa.

Africa is, even so, awash with rebel insurgencies and social conflicts, many of which have an impact on oil. In Chad, Muslim and ethnic separatists have added fuel to the fires of the north–south, Muslim–Christian divide that runs across the Sahelian region. With the opening up of the Doba Basin, funds from bonuses were diverted to the military against the wishes of the IBRD. A carefully crafted paternalist oil management regime was constructed, allowing ExxonMobil's venture (with Petronas and Chevron) to proceed. The Chad government's manoeuvres to modify this treaty, even temporarily expelling Chevron and Petronas in 2006, reflect the deep interest of its minority political elite to assert enhanced control over oil revenues. And it is not certain that the issue will not be reopened, given that it presents a challenge to African sovereignty. The spillage of conflict from Darfur into eastern Chad has added new instabilities to a country that long before oil has suffered from irredentist movements. Much of the drive for insurgency stems from contested state legitimacy, and reflects power struggles within the barbarian world.

In Nigeria, *sharia*-inclined Muslims from the northern-based Hausa and Fulani have clashed violently on many occasions with Christians and the oil-rich south, and with the Yoruba and numerous other ethnic groups populating the oil-rich delta region. Over the years many communities have become engaged in low-level conflict, now much aggravated, with the government. In these circumstances the oil companies (including, notably, Shell and Chevron) face organised struggle, including from the Movement for the Emancipation of the Niger Delta (MEND) and others over oil rights and control in the poverty-stricken delta.

While oil is certainly contentious, and has become the heart of the matter, many separate social, ethnic and political conflicts mark this landscape of turmoil.

Cameroon too has exhibited social division, based on religious and ethnic grounds, although the long-standing conflict with Nigeria over the Bakassi Peninsula has finally been settled. At the time of construction the Chad–Cameroon oil pipeline raised new concerns about "disadvantaged communities" lacking political leverage and influence. In Liberia, bloody wars have been the rule until only recently. This tradition was continued by the conflict between former President Charles Taylor and rebel forces until Taylor was ousted. The struggle destabilised Guinea, Sierra Leone and parts of the interior ethnic areas beyond the direct control of the Liberian government. All this affected offshore acreage leasings, eventually concluded, in both Sierra Leone and Liberia. In Senegal, the southern zone of Casamance has long been subject to a dispute over autonomy sought by the MFDC (Mouvement des Forces Démocratiques de la Casamance). This may deter some players from taking up positions onshore. Throughout these states historical antecedents to oil have been a prominent feature of the political and military struggles.

The Maghreb region has been far from insulated from the types of conflicts and instabilities that have permeated Sub-Saharan Africa. Ever since Nasser, the Muslim Brotherhood has been at odds with Egypt's secular government, with episodic civil conflict the result. Amid deteriorating social conditions and the rise of Al-Qaeda, the rise of Egyptian Islamic *jihad* has added global dimensions to the struggle. Issues of state control, human rights and poverty have played a large part, and hydrocarbons interests have not been the primary driver of contestation.

In Morocco, the Polisario (the armed wing of the SADR) remains a thorn in the side of the government. The zone of southern Morocco (depicted by antagonists as Western Sahara) is still disputed. Offshore rights there have been awarded both by the government and, more recently, by the SADR to competitive companies unable to exercise them. Total and Kerr-McGee decided to exit this play, though Kosmos remains in place and new entrants in southern Morocco are expected soon. Mauritania has not been immune from coup initiatives, with attempts in 2003 and then a coup in 2005 that installed a new military government, an event that later resulted in elections in 2007. This had implications for offshore players such as Woodside Energy at the time, forcing the renegotiation of some contracts. Oil has played a role in all these arenas and yet deeper social conflicts underlie and predate oil as prime cause.

In the 1970–80s, Eritrea and Ethiopia fought a long internecine, Cold War-induced conflict with substantial casualties before the fall of the Ethiopian president, Haile Miriam Mengistu. These states fought each other again in 1998–2000 at a cost of 50,000 casualties over land demarcations, since when there has been a state of uneasy tension while the theatre of conflict and proxy war has switched to Somalia. Ethiopia has been able to attract a few oil investors, notably Petronas and some smaller independents. Only a few players (such as Lundin Oil), however, have been prepared to invest in Eritrea. To some extent within the Horn of Africa much of the turmoil has related to unsettled questions of regional hegemony, the Ethiopian government claiming this historic status and becoming aggrieved when others challenge the boundaries and presumed dominance of its ancient empire.

In the Democratic Republic of Congo (DRC), conflict over control of the Great Lakes during 1999–2004 attracted the national armies of Zimbabwe, Angola, Namibia, Rwanda and Uganda as well as numerous rebel interests. Oil was not directly the issue. The conflict did, however, delay the opening up of African Rift plays. A new era is now promised, though recent peace agreements remain to be confirmed in practice, and even Kinshasha saw fighting in 2007. As an extension of the exploration of the Albertine Graben zone in Uganda, only small independents such as Tullow Oil and Heritage Oil have entered the DRC. Heritage's original deal was negotiated in part with DRC rebels and was later called into question. Although some of the rebels have joined a coalition government in Kinshasa, many remain in the field and armed. This unstable situation hinders the full opening up of the central interior basins in the DRC's vast Cuvette zone. Meanwhile, Uganda has experienced internal military struggles involving both the Lord's Resistance Army and the Acholi ethnic groups. It remains to be seen whether the recent discovery of oil might produce new tensions.

Overall Africa has witnessed significant conflict over the decades, much unrelated in its origins to oil questions – but over time the petroleum bounty, now much augmented from discovery and development, has entered the equation. In future the crude stakes may become more prominent, and ambitions around oil may well lubricate further tensions and conflict as African states mature and go through uneven processes of internal realignment in social, economic and political order. Perhaps the lesson here has been that as old imperiums retreat, and empires fall, new empires of oil and power seek to enter the vacuum created. All this has been fertile ground for the new scramble for Africa – by Africans, with and without allies drawn from many different oil worlds abroad.

❋

In Latin America there have been many social conflicts still without resolution – most notably in Colombia, but also in Mexico where the Zapatistas and Mayan Indians in Chiapas have fought for autonomy. Although many of the continent's conflicts have been intrastate, they have also on occasion involved adjacent territories – often with an impact on the oil industry. In Colombia the struggle with FARC (Fuerzas Armadas Revolucionarias de Colombia) has lasted for four decades. In the process it has tested the viability of the oil industry's operations and pipeline security as well as the legitimacy of the Colombian state to control exploration and development. The opposition of FARC and the Cuban-inspired ELN (Ejército de Liberación Nacional) to the exploitation of the Colombian oil resource by corporate oil has been at the heart of this conflict. Recent US military support for the Colombian government has been directed in part at protecting US corporate oil interests in Colombia. Yet complex non-oil considerations have also been found over the years embroiled inside this conflict, including competing political ideologies, indigenous separatism, financial power and narcotics. Cold War legacies have also played a part.

In Peru, the Shining Path and other rebels have fought the government for political and ideological control of the interior basins. This hampered the exploitation of the Oriente hydrocarbon play. There has also been some overflow from the Colombian conflict in the form of activity by FARC rebels in adjacent states. As elsewhere in the Andean belt, the rise of indigenous peoples against once-established, post-colonial Spanish dominance has been central to much conflict.

Meanwhile, in Guatemala the government has been in armed conflict with Mayans for four decades. A full settlement has yet to be reached. This consideration has influenced some oil company assessments of the desirability of acreage-leasing and investment in the country. For countries such as this – not central to world powers or empires of oil in contestation – local issues inevitably come to the fore.

Latin America is now in the throes of new uncertainties, even social turmoil, related to indigenous rights, contestation with the elite and control over the levers of power. The Bolivarian revolution promoted by Chavez has added new dimensions and has had regional impacts in Bolivia and Ecuador, and long-unsettled local issues have come critical. The oil and gas resources now form a key part of this mix along with issues connected to inequality and asset ownership, but petroleum has not been the only element. The new political regimes being installed in large swathes of Latin America probably represent equally the forces of de-linking in this hemisphere from its earlier anchors tied to once-unquestioned American dominance.

✳

In Asia, too, long-standing conflicts have affected the oil industry. In Myanmar, Shan separatists and Muslim extremists have contested power with the junta since the days of the SLORC (State Law and Order Restoration Council). Myanmar has also engaged in border skirmishes with Thailand, even while the state oil company PTTEP remained a major gas investor in Myanmar's gas-rich offshore. Myanmar remains under sanctions. Several US players have abandoned interests there (notably Texaco and Arco), leaving Chevron in place after its acquisition of Unocal's assets. Issues of democracy and military rule have predominated, but it would be unwise to ignore the impact of hydrocarbons on the intent of the state to resist change. Even so, local issues will clash with a wider set of oil and gas interests – not just those of Indochina, but now including Chinese, Indian and Korean players.

In Indonesia, the archipelago has witnessed a range of conflicts related to regional autonomy, ethnicity and religion. Islamic pressures have grown over time following the fall of the Soeharto government. Also at issue has been government control of oil resources in and around Aceh, West Papua, Kalimantan, Sulawesi and Ambon. The Aceh struggle has broadly been settled, with the state ceding influence in oil matters to the provinces and so complicating relations between companies and the power structure. Instability in Indonesia has had many roots, with regionalism, poverty, state management, ethnic balances and economic strategy among them. In many respects, under the 25-year-long Soeharto regime's unifying Pancasila philosophy, Indonesia saw itself as an empire alone, even the main driver within ASEAN – a role much reduced since the glue of the government's centralist grip has loosened.

China has now asserted a new prominence inside Asia not there before, and is a key player as an empire of oil in the world at large. It has shifted focus from its long border with Russia to the South China Sea, where ownership of the Spratleys and Paracels – key issues for the many states claiming offshore rights – remains unresolved. Jurisdiction in this region is contested among China, Indonesia, Malaysia, Vietnam, Brunei, Taiwan and the Philippines. It is a vast area in which much is still unknown about oil potential although China considers the zone highly prospective. In its far west, China has taken military measures to defend Xinjiang against Muslim Uighur separatists and to secure zonal control and gas pipelines to the eastern coastal markets. It has elevated its conflict against the East Turkmen Islamic Movement to the status of a war on terror and has increased the number of Han Chinese in Muslim Turkic-speaking zones. Issues of autonomy, conflict and repression have been overlain with the needs to exploit the Tarim Basin and to control gas flows to coastal China from interior basins and even, eventually, from Kazakhstan and Turkmenistan. In almost all respects, China can be seen to be exercising an imperial tutelage – seeking state-driven hydrocarbon access – both inside and outside around the boundaries of the once-Maoist state.

In South Asia there has been on–off conflict between India and Pakistan since partition in 1948, continuing to this day in Kashmir despite recent initiatives. Though this has little direct bearing on oil, it does have huge ramifications for Iran–India gas links, cross-border energy deals and the provision of Turkmen gas to Pakistan and India. Within Pakistan, the central government's influence has been much restricted in areas such as Baluchistan, the Afghan border and the tribal areas of Waziristan. In these areas, now much infiltrated by the Taliban and Al-Qaeda, the potential for instability has had a negative impact on the oil game. With turmoil not uncommon in Pakistan and accentuated in 2007, India's regional energy strategy looks more uncertain. Its global oil search, mimicking China's, now has a global remit as its state and even private companies have taken to pastures afar – in Africa, the Middle East, Russia, Australia, Canada and Latin America.

In the 1990s Central Asia developed into a new theatre of conflict as the West became keen both to secure access to reserves and to ensure that oil supplies would be routed west through Turkey rather than through Iran. This formed part of a global diversification strategy to limit dependence on the Gulf. The US government has made little secret of its attempts to woo the region's post-Soviet leaderships, regardless of the latter's author-itarianism. Now US military networks have been installed to protect the future of the region from potential irredentism, Islamic fundamentalism, terrorism, Russian recidivism and potential encroachment from Iran. In America's eyes, continuing Western investment in the Caspian strengthens the rationale for such installations. But in 2007 it was Russia that secured agreement for Turkmen gas pipelines to go though its territories, outflanking America's preference for gas flows to traverse the Caspian and head towards Turkey.

Russian military strategy has also been reoriented in recent years. Disputes over the Caspian Sea remain unresolved, and gunboat diplomacy has been exercised at times while negotiations have continued over the contested oil prize. Around the Caspian zone, Turkey and Armenia hover with strategic interests in mind. Russia is concerned about the potential for Islamic insurgency in the region, especially in Uzbekistan and Kyrgyzstan, which are crucial for the transshipment of oil supplies. Georgia's independence has been greatly resented. Concerns over gas supplies have helped to drive current disputes and military actions related to the self-styled republics of Abkhazia and South Ossetia. Russia retains a network of military bases in southern Russia and the Caspian region, with allied military contacts, arms deals and strategic agreements. The Russian bear is unwilling to abandon all its cubs that were once closely held.

Russia and Chechnya have conducted a second war over Grozny's quest for Muslim-led autonomy. Russia's opposition to a fundamentalist regime, laced with Wahhabi-based Islam, has been brutal. The conflict has become linked to both the war on terror and regional oil issues. It has stunted the development of the Chechnyan oilfields. The desire to secure and protect the Georgian transit route for the Baku–Tbilisi–Ceyhan (BTC) oil pipeline helps partly to explain America's reluctance to pressurise Moscow on concerns over human rights that have arisen in this conflict. Meanwhile, in Azerbaijan war has shaped relationships with Armenia and the enclave of Nagorno-Karabakh. With the rise of Azerbaijan as a large net producer and growing oil exporter, new concerns have translated into issues of revenue management and the internal allocation of funds controlled by an increasingly oil-rich but authoritarian state.

From the rapid dissolution of the Soviet empire has emerged a new scramble for oil involving the great powers both directly and indirectly. The BTC pipeline was not just an oil deal: it was a political and business strategy set within a security framework and military engagement. Oil and gas pipeline logic has not always prevailed over politics. Unocal's CentGas project (with Delta, Inpex, Itochu and Crescent) for exporting gas by pipeline from Turkmenistan through Afghanistan, previously agreed with the Afghan government, fell foul of the Taliban in 1998. This venture incorporated a 1 MMBOPD oil pipeline along similar routes that would have backed into the Russian pipeline system. This too is now in abeyance, though it might re-emerge if Afghanistan were ever to be properly stabilised.

Iran is considered by many global strategists in the US as a growing threat, not just in the oil game but in its sponsorship of terrorism, and was listed as part of the "axis of evil". It has deployed its growing naval capacity in the Gulf (whence over 18 MMBOPD flow to Western markets). Iran could prove a thorn in the side of the Caspian powers and their oil companies if it wished. It has made moves to control islands in the Gulf, and stands in potential conflict with the UAE over several such areas. Iran's growing ability to generate nuclear energy allied to alleged ambition to develop nuclear weaponry, its theocratic regime, and relationships both with fundamentalists and armed groups such as Hamas and Hizbullah, ensure that the country is perceived as a central threat to US and Western diplomatic and military strategy in the region. It is by no means clear how this confrontation between America and a hydrocarbon-dependent Iran will finally play out.

The build-up of military power continues throughout the Middle East, the world's central bank of oil potential. The pre-positioning of forces, including rapid deployment forces, blue navy capacity and essential military stocks, has occurred already in the Gulf and Saudi Arabia, Bahrain, Kuwait and Qatar. The strengthening of American forces in Qatar is not purely the result of the demands of the war in Iraq: that Qatar has become a senior regional command post is also the result of strategic diversification, involving less reliance on Saudi Arabia. These moves may be read as purely oil-related, but they equally reflect great-power competition around the southern flanks of Eurasia, and the desire on the part of many contestants for regional hegemony and local influence.

✳

Across the world, there have been numerous zonal conflicts over boundaries that sit atop known or prognosticated oil and gas fields both onshore and offshore. Many have been settled (often with joint development area arrangements). They include agreements made between Malaysia and Thailand, Australia and East Timor, Nigeria and São Tomé & Príncipe, and others. Many boundary conflicts, however, remain unresolved. Some result from particularly fractious states or polities (for example, offshore southern Morocco), antipathies between leaders (for instance, Gabon and Equatorial Guinea), poorly designed national boundaries inherited from past empires, or social complexities that cannot readily be settled in juridical forums. EEZ claims may in time add to this list. A world without contesting states, in which potential oil resources are at stake, appears unlikely for the future.

It is not only oil and gas reserves that feature in security concerns and military strategy. The same is true of pipelines and other routes for transporting oil and gas. Many long-standing ethnic–corporate conflicts have been shaped around oil pipelines in the Niger Delta. In Sudan, security of the pipeline to Port Sudan proved a key issue in enabling exports to flow. The Caspian oil pipeline route selection was one of the most fractious in recent geopolitics. The "Velvet Revolution" in Tbilisi in 2003 was monitored closely by the US for its implications for the pipelines traversing Georgia. Political change in Ukraine also caused anxiety, as elsewhere in the east European transit zones connecting gas to the EU, with Russian strategy shifting to tougher management approaches in gas supplies and pricing through Gazprom. This game is likely to continue.

Moreover, many global choke-points in the oil business remain wedged between states with divergent political, ideo-logical, religious and ethnic positions – notably the Strait of Hormuz, the Persian Gulf, the long sea-lanes to Japan and China, the Suez Canal, the Red Sea, the Malacca Straits and the Bospho-rus–Turkish Straits. None of these lies directly within American, Russian, or Chinese control. Around 30 MMBOPD (nearly 40% of global daily demand) pass through these choke-points. Any threat to transshipment could serve as the trigger for conflict between the dominant empires of power, as was the case when the Suez Canal was closed under Nasser.

Much of America's strategy has been driven by its need to manage and reduce its current dependence on the Middle East and to diversify its supplies of oil in order to lessen that dependence. America already imports 12 MMBOPD worldwide, and this volume is projected to rise to 20 MMBOPD by 2025. In the circumstances, the possibility of Iraq under Saddam Hussein developing nuclear ambitions (or even other weapons of mass destruction) was not one that America would accept. The attempt to exert military control involved a calculus typical of empire, and Iraq's oil potential could not have been unnoticed.

The US has not yet been prepared to back sufficient future options as ultimate hedges to deepening oil dependence and foreign crude imports – such as the eventual large-scale emergence of a hybrid of domestic synfuels, shale oil, tar sands, biofuels, or a hydrogen-based economy. Yet it is heading in that direction. The rapid development of imported liquefied natural gas (LNG) is certainly part of America's strategic response, but it is one that provides only limited security protection in energy. America is therefore likely to pursue further its strategy to diversify oil supply. Its focus has been on Africa and Latin America, though in the latter case the situation has been greatly complicated by the coming to power of Hugo Chavez in Venezuela. The strategy depends on a number of factors, including stability and continuity in Colombia and Nigeria, growth in Angolan production and new supplies from the Gulf of Guinea. In all domains competition from China – especially as a crude oil offtaker – has emerged.

✳

For corporate oil, questions about the cause, incidence and probability of war are far from academic. The future does not look conflict-free, and may never have done so. But unique circumstances now prevail. Companies will thus need to plan with an expectation of turbulence and widespread instability. Earlier benign scenarios, based on "peace dividend" prosperity and the rapid diffusion of capitalist democracy, which flourished at the end of the Cold War, have disappeared. At best, the new world order may be said to contain elements of "structured disorder". The industry will need the best possible alarm systems for anticipating conflicts and even wars, especially where these might affect core portfolio.

One oft-encountered set of conflicts will be those associated with failed states. The likely incidence and distribution of these may alter from the current collage. Trends from the past cannot be assumed to continue in linear progression. Consider Africa. Though there have been exceptions (notably the DRC), many of the conflicts there over the past couple of decades have arisen among the smaller states – Liberia, Sierra Leone, Burundi and Somalia, for example. Now, however, many of the larger polities, including Côte d'Ivoire, Zimbabwe, perhaps in future Nigeria, and others are at risk. Of the greatest importance is the question mark over the continued integrity of Nigeria. Central authority has been in question, with the social order fraying at the margins among growing numbers within disadvantaged and disenfranchised communities, ethnic and Muslim–Christian division at its deepest, and oil stakes at their highest. It was after all the Biafran secessionist war (1967–70), fought against the Igbo in and on top of the delta's oilfields, which nearly tore Nigeria asunder.

Nobody reading the survey of conflict provided here could doubt that oil and war have often been interrelated, if not primal cause at the outset then in relation to the stakes involved in the outcome. The frequency and serial continuity of intrastate and interstate wars has been a mark of the political landscape in modern human history. Oil has been an associated or causal factor in numerous wars. War in turn has in many places had a severe impact on the oil game. But it has been a complex relationship, not a unidirectional one.

Therefore it is not the case that we can assume an inevitable nexus between oil and war, or a clash between empires of oil that must lead to war. Man's pugnacity, after all, predates the arrival of hydrocarbons and wider interest in oil. It has been fed by many causes, among them political ambition, demagoguery, predatory behaviour over territory and non-oil resources, disputes over ethnic, religious and nationalist issues, or claims over autonomy and governmental authority.

Indeed, many states have managed to avoid overt and violent conflict over their reserves of oil and gas. Norway, the UK (notwithstanding Scottish angst), Malaysia, Brazil, Brunei, Argentina, Canada have been notable examples. In these instances the core issue has been over national and regional interest. Moreover, to say that oil attracts armed conflict, even war, is not always to say that it alone is the cause of war, or that oil causes war, which often hinges on multiple and complex historical provocations, economic concerns, political dramas, military incidents, ethnic feuds and ancient unresolved rivalries.

Consider, for example, the association between hydrocarbons

and wars in Afghanistan. The past plans of Unocal, Bridas and partner-investors in Turkmenistan to export gas to Pakistan and elsewhere were unsettled by the Taliban regime and comprom-ised for many years, both under Soviet tutelage and afterwards. Warlordism, the fragility of coalition government and the resur-gence of the Taliban mean that the venture entails high long-term risk. The Afghan serial conflict itself, however, cannot be said to have resulted purely from hydrocarbon considerations. There has been extensive interplay with multiple factors: enmities and insur-gencies, Russia's history of intervention, the role of narcotics, ethnic divides, the record of the Taliban regime, the presence of Al-Qaeda and more traditional forms of terror and violence. For all intents and purposes, Afghanistan's small oil and gas fields have been a minor consideration.

It would be foolish nonetheless to deny the possibility of further oil-related wars. The potential triggers are numerous. Yet equally it cannot be assumed that such wars are inevitably inscribed into world oil futures. Conditions, after all, may change in unexpected directions, sometimes even reverse. A radical fall in the price of crude, for example, would devalue oil assets and flood markets, removing some of the fractious conditions that many comment-ators evoking oil-determinism assume will continue. Similarly, there is room for co-operation between state oil companies and corporate oil, as well as benign competition. Not all these streets are one-way.

The conviction that the oil/war nexus is unavoidable perhaps stems directly from the liberal ideologies of many doomsayers. Sympathy for the underdog, the wretched of the earth, is entirely understandable. And it is always easier to blame single causes than grapple with historic complexity. Therefore, the knee-jerk tendency of many, and many misinformed, to see the oil game in general – and corporate oil in particular – as the archetypal root of all evil is understandable. At best this is a poor caricature of reality. But it is necessary only to consider how Machiavelli would view the foundations of such claims to realise that they are woefully inadequate and a danger to clarity in analysis, a fact that should not be overlooked.

✳

The conception of oil as a curse rests on its association not only with so-called oil wars, but also with poverty (whether in the sense of low living standards or related to "relative" poverty, that is, derived from economic inequality). It is not difficult to find states that combine considerable oil endowment with widespread poverty in this sense. They include, for example, Angola, Nigeria, Saudi Arabia, many Central Asian states, Indonesia, Venezuela, Colombia and Equatorial Guinea, to name a few. It is equally possible to find many states without hydrocarbons having similar or even more acute conditions of poverty, say in Africa (such as Malawi, Lesotho, Zambia) or elsewhere (Nepal, Bhutan).

As with war, however, the correlation between oil and poverty is far from perfect. No one would doubt the existence of poverty in many pre-oil societies. There are many causes of impoverishment as the literature on development economics instructs – including economic mismanagement and in some cases war. Oil has become associated with poverty in part because of its sometime association with the latter. Many non-oil societies today exhibit extreme levels of poverty – Ethiopia, Eritrea, Cambodia and Laos, for example – while many states lacking oil endowment have become prosperous (Japan, Singapore, Botswana and Mauritius, for instance). This taxonomy invokes gross simplifications – just like the oil/poverty nexus so widely alleged worldwide.

The debate on world poverty and inequality is also one that is contentious. The UN has often identified growing poverty in the world. Some recent studies suggest that inequality, having risen for 200 years, reached a plateau from 1980 onwards. The numbers living in extreme poverty (living on less than $1 a day) are estimated to have fallen mainly because of fast growth in China and India. Globalisation may, it seems, have had some benign aspects. But less of this poverty-alleviation appears to have been in Africa. Many contradictory data and calculations exist on this vexed question.

Whatever the global trends over time, however, we can consider the petro-states in particular even if measurement difficulties are profound. The evidence and sight of oil-created wealth in a sea of absolute and/or relative poverty is not uncommon: Angola is an example. Inequalities of income in developing worlds are typically vast – those of wealth, even larger – and more pronounced than in rich worlds. Within many petro-states as a whole, income/ wealth gaps have probably widened. It is as if some states (but not Norway) have shed their systems of traditional or organised welfare based on historical patterns in exchange for what some might describe as the "new extremism" – that is, islands of pros- perous oil modernity amid a sea of traditional deprivation.

This dilemma poses serious problems for corporate oil. For many observers, looking on at the juxtaposition of oil, wealthy elites and shantytown wretchedness, the facts "speak for themselves". That is, the juxtaposition is taken as evidence of causation. Evidently for some, corporate oil is taken as the source of all economic woes.

There are, of course, ways of accounting for the juxtaposition. The concentration of poverty and the depth of inequality are often at their greatest, after all, in societies that have failed, suffered aggravated conflict or war, or are closed and autarkic, run by autocratic or feudal regimes. A smorgasbord of complex reasons may be causal. Corporate oil's global search for oil and gas, especially in the age of portfolio diversification, of necessity takes more companies into many such poor societies. In the process, it seems, it has become guilty by association.

But – regardless of the rights and wrongs of the arguments – the brute fact that corporate oil has to deal with is that, amid conditions of poverty and inequality, it is frequently nominated as the guilty party. Oil companies, as the bridgeheads of the rich world, are often seen or depicted as the agents of exploitation – the global oil robber barons. They thereby have often become targets in their own right.

Companies become targets too because they are associated with perceived imperialism. Oil companies are many a time regarded as the "lackeys and cohorts" of American empire or Western interests. Suitable distinctions are not always observed: even non-US western companies are labelled so. Yet at times foreign state oil companies found in the same milieu are given a free pass on this process of tainting. Targeting can be dangerous and may bring with it the threat of violence. Oil and poverty have proved a lethal cocktail in, for example, the Niger Delta and Colombia. It has often been observed that some petro-states, even states without oil, have been the breeding grounds of inter-national terrorism. The implied association of oil and exploitation with poverty creates a veritable tinderbox, even if corporate oil has not been causal to the basic problem.

Part 2

EMPIRES OF OIL

The fate of corporate oil is intimately bound up with the nature, strategy and international role of America as the leading empire of the Western world. At issue are questions of how the US articulates its growing oil and gas/LNG interests drawn from home zones and the wider world, the political and energy leadership it provides, and the reactions it produces among related powers. Within America there has been much debate over the extent to which it is, or should be, an imperial power. Here we examine the geopolitical landscape of the world in general and, centrally, the role of the American empire. We consider the current reality and potential future of this Colossus and the implications for global oil and energy strategy.

The end of the Cold War left America as the pre-eminent power in a changed world, standing head and shoulders above even such considerable powers as resurgent China, fast-modernising India and a much-shrunken Russia under reconstruction. This had important consequences for the world oil industry. Whatever America decided was of cardinal importance, not only for US corporate oil but also for oil companies globally. For a period in the 1990s this situation looked, at least to Western eyes, stable and durable. Many commentators unwisely rushed to judgement and forecast that it would be bolstered by a global diffusion of democratic capitalism and the triumph of liberal philosophy.

Clearly democratic capitalism has not become the global norm. *The Economist* reports that, while according to Freedom House there were in early 2007 up to 122 electoral democracies (64% of all states, compared with 40% in 1980), in fact only 46% (89 states) were "politically free". It adds that the push towards democracy has stalled if not reversed.[57] *The Economist*'s own democratic index records only 28 full democracies, with a further 54 flawed ones, as well as 30 hybrid regimes and 55 authoritarian states. The democracies that do exist are not guaranteed to survive, especially in the developing world where order seems more often likely to produce democracy than vice versa.

The world, then, did not develop according to any preordained plan. Now the developments of 11 September 2001, US military action in Iraq and the emergence of rival empires has radically changed the geopolitical scene. The US now has a "watch list" of 25 countries in which instability might precipitate intervention. Looking at the world as a whole, and abstracting from its official cartography, we are left with the impression of a melange consisting of modernity in places, emergent economies and fragile states in others, with quasi-city states dominant in some arenas amid a kaleidoscope of ancient cultures with primitive or pre-modern conditions still widespread. This is an inherently unequal, unstable global system. For all the shocks and challenges, however, America's empire of oil has to date adjusted or in places (in the face of resource nationalism) eroded, rather than simply imploded.

The fractured globalisation of the contemporary world presents corporate oil with a different challenge from the benign globalisation once forecast – one that takes us into the sphere of international relations, diplomacy and warfare. In this context the Roman Empire/Barbarian template appears more apposite and in tune with emerging-world geopolitics than that of any liberalising, neutral, open market model. Whatever Western precepts might tempt us to believe, the barbarians outside the empire are not lacking in strategies, skills, political and commercial knowledge, localised military capabilities, or moral fibre with which to secure and protect their strategic oil interests. Theirs is a world that has already experienced classic imperialism and, in many cases, modern military interventions. Lessons have been learnt.

The Roman Empire dealt with problematic conditions not so dissimilar to those found today. It too had its rise, peak, plateau and eventual decline, and faced turbulent times. Some analyses have connected its ultimate demise to its affluence and decadence counterpoised against what were in the end much tougher barbarian cultures. The Roman Empire yielded to an overstretching of resources combined with inner decay, rivalry and an unwilling-ness to engage in the necessities of empire consolidation.

In the end, Rome's social construction built on its legal system was not enough to sustain its perpetual continuity. In the 21st century, it is oil (or more pertinently sufficient affordable energy) that is essential if the American empire is to sustain itself. Yet so too is it for the emerging empires. As a result, it is the global

97

art of strategy that is now critical to America and corporate oil. According to Sun-Tzu in the ancient classic *The Art of Warfare*, the real winner is the country that can avoid war or can select battles carefully, with judgements based on deep appreciation of best strategy. It does not seem that America has yet learned all these lessons.

Corporate oil may decide to think similarly. It may need, in particular, to develop much better counter-intelligence on world conditions and to develop new alliances and managed relationships to guide it through the labyrinthine social disorders now littering the barbarian oil world. It may do corporate oil little good to adopt a sanctimonious attitude, presume its affluence is guaranteed, substitute public relations emotionalism and good intentions for strategy substance, or remain unaware of anarchic potential while appeasing hostile localised power. Many of the conditions encountered by corporate oil ventures are far from mirror-images of its home base.

Oil access for companies is now the key issue, especially in view of the challenge from resource nationalism. The richer modern world, with corporate oil in a position of relative affluence, is tempted to rely on strategies based on big-bang fixes and altruism and to envisage world change as a morality play. Given the prevalence of self-interest in oil matters and the increasing need to maintain operations in conditions of unrest or conflict, the need for realism is paramount. At present, corporate oil's barbarian opponents appear savvier about conditions on the ground than are the senior corporate managers sitting in the safe, well-ordered world of the executive suite.

However, the oil world is strangely self-censoring and in many cases self-effacing. As shown in Chapter 9 (and also, in more detail, in Part 3), corporate oil is reluctant to tackle NGOs head-on in relation to many contentious issues. As a result, it conveys only a muted message to the wider public and surrenders much of the strategic terrain to its antagonists. Putting this policy into reverse will now not be easy. But it may become necessary.

Legal restraints and self-elected sensitivities may even prove to be among corporate oil's greatest vulnerabilities. The oil world is now imbued with fear of liability for implied or collateral damage, concern for environmental risks, concern about potential social costs and worry over a minefield of "ethical practices". Corporate oil's enemies for the most part do not face the same restraints. They are far less accountable for their actions. They can more often operate from the premise of moral absolutism. They make their case worldwide in a critique that often remains unanswered and that ties corporate oil to the alleged sins of the empire of oil.

5

The American empire

The sheer scale of the American empire is awesome. It brooks no comparison with any historical antecedent – Spanish, Portuguese, Austro-Hungarian, Ottoman, British, French or Soviet. In terms of military capacity, the US enjoys nuclear superiority, air power with a global reach and five-ocean naval dominance. It spends more on its military and global defence than any other country – indeed, more than the next 15 countries combined. Equally, it still enjoys a global economic superpower status. Notwithstanding some weakness in recent years, with American GDP shrinking in relative terms on a global scale, and transition to the euro mooted in some oil deals, the dollar remains the world's dominant currency. America's hyper-puissance extends too into some areas of "soft power" – the cultural strengths provided by the reach and diffusion of American images, icons and brands. Corporate oil has traded off the sense of superiority this provides, including technology prowess.

Yet the lessons of long-term history suggest that America's superiority is far from guaranteed to last. Unipolar world order has disappeared. On each front, America faces new challenges.

The dollar could be vulnerable to a switch by Chinese and Asian bond and US asset holders and Venezuela and Iran have already expressed interest in euro-denominated oil trades. American iconography is the subject of critique and control, especially in the Islamic world. In geopolitics, America confronts a new decades-long Great Game with current and emerging great powers, notably Russia and China. Whereas, in the final decade of the 20th century, many pundits in America assured themselves that no global conflict or challenge over the coming quarter of a century was likely, now the prospect of an American apogee is widely discussed.

A similar picture obtains in the oil world. America's predominance in corporate oil has ceded partial ground to the challenge of Europe's super-majors (BP, Shell, Total, ENI, Repsol-YPF). Many global independents can no longer be said to have US domicile and significant players have emerged on all continents. National oil companies account for a growing share of upstream reserves, deals and activity, and they are increasingly international, and some global, in their operations. Former American players (Amoco) and independents (Arco) have been bought by foreign players. America's sanctioning of many states has by default given less constrained foreign companies some competitive advantage. In addition, many countries besides America now possess advanced technology in such areas as deepwater oil, geoscience, LNG-GTL and exploration – for instance, Norway and StatoilHydro. Crucially, too, many have deepening strategic oil relationships with powerful state players – this is no longer an American preserve. US "soft power" in the diplomatic oil game is no longer unique.

On a day-to-day level, business might seem to be proceeding normally for American corporate oil. It enjoys residual strengths. Yet the umbrella of Pax Americana under which it operates is weaker than before. Strategic oil competitors abound from the top of the corporate food chain to the bottom feeders proliferating in the long corporate oil tail that has emerged across the world. The likelihood is that the shift will continue under pressure from localisation and new company formation inside the developing world, the growing internationalisation of national oil companies, a new set of global portfolio strategies and alliances struck, and the worldwide search for equity crude. Petro-China, CNOOC, Petrobras, ONGC, Petronas, KNOC, Gazprom, Lukoil, ENI, Sonatrach, StatoilHydro and many other state entities are enlarged and aggressive players in this competitive game.

Corporate oil today has to operate in a more complex world than ever before. In the 19th century, a handful of great powers managed a system of 40 or so states, as well as many closely held colonies. From the turmoil of the 20th century, including the post-war surge of decolonisation, emerged a bipolar system amid approximately 70 states. This in turn yielded at the end of the Cold War to a fleeting period of unipolar prominence. Today, however, when the oil industry is more globalised than ever before, there are nearly 200 "official" states with many de facto regimes and statelets inside and around them. It is conceivable that further balkanisation could occur over the coming decades in Africa or Asia, so that the number of states (recognised or not) and segmented jurisdictional zones (tolerated or otherwise) might rise still further.

In contrast to the clearer-cut identities of the Cold War, the formal and informal enemies of the American empire today are often obscure or opaque. National boundaries and the power of the nation state have diminished in importance. The exercise of state authority has become more "lateral" and, in some arenas, almost chaotic. The new oppositions to American hegemony, defined in religious, ethnic and cultural terms, reflect shifting quasi-political formations. It is a game for power and control played in the shadows and often underground. These are terrains to which corporate oil is often unaccustomed.

Some argue that, in the long run, the reserve replenishment rates needed for corporate oil to meet its targets for sustained production growth will require either the dissolution of OPEC or a radical change in its policies on access. OPEC's hand on world oil supply and markets may not be the main issue. It is rather that the club (as OPEC's key states may be known) acts as a mechanism to thwart American corporate oil's access to huge proven oil reserves. The invasion of Iraq may have been expected to ameliorate this situation by providing new leasings – a prospect abhorred by hardcore OPEC players. Indeed, America's real-politik in recent strategy – including the de-sanctioning of Libya, Sudanese rapprochement and private encouragement by some in America for Nigeria to leave OPEC – may be read at the level of grand strategy as largely access-driven.

The debate on America's global strategy has covered a wide range of possible positions, including isolationism, unilateralism, multilateralism and the need for wider allied coalitions. Central to the debate has been the question of the extent and desirability of imperial power. Critics, both secular-radical and Islamic, accord oil a central role in such debate. They are inclined to see the American empire as purely materialistic and its foreign policy, therefore, as little more than a means to acquire the oil and energy supplies that this requires. Some regard America's materialistic imperialism as the "Mother of all Fundamentalisms".

The Islamic world sits atop many of the world's major oil reserves: hence, say the critics, its targeting by America. The portrayal of Islam as anti-modern and (from a Christian fundamentalist point of view) somehow ungodly provides convenient cover in the guise of a modern crusade. The war on terror is thereby viewed as nothing more than a war to secure oil. America's opponents, so the critics argue, are demonised and war created according to the interests of oil – as witnessed by the transformation of Saddam Hussein from a Western ally to a leader in the famous axis of evil. International institutions such as the IMF, add these critics, exist only to do America's bidding by other means.

Against a polemical background of this kind, corporate oil faces a difficult challenge in its attempt to manage public relations and influence world debate. Such discourse is not quaintly confined to the drawing rooms of Western capital cities and the international bourgeoisie. It is wired worldwide through the internet, and into the cities of the developing world and onto the Arab streets. Corporate oil's reach is limited in these latter arenas.

American corporate oil faces the difficulty that many of its political opponents are states that the industry may wish or need to do business with now and in future. Meanwhile, most competitors (some allies included) have been prepared for their oil companies to plough their own furrows unhindered. France does not inhibit Total's dalliance with sanctioned oil states. China has taken up oil positions through state companies in contentious countries and projects. Russian oil companies have been active in Iran and Sudan, and they want access to Iraq. How long will it be before these environments become open for American corporate oil entry? Already the winds of change have blown in Libya, where for over a decade former European state oil entities (Total, ENI, Hydro and OMV) operated untouched in US-sanctioned Libya. It cannot be in the long-term interest of American players to remain permanently locked out of sanctioned oil-rich territories or even OPEC worlds around which the shutters have been raised.

In this Great Oil Game around world oil reserve access, so different from the Russo-British 19th-century version in Central Asia, there is the deeply embedded temptation for empires to protect their own interests (oil, standards of living and lifestyle).[58] However, US reach in the oil world is more limited than its desires. It would be surprising if after the occupation Iraq did not revert to a policy of open access to companies worldwide, including foreign state players, for diversity in investment as well as for historic-political reasons – this is indicated already in the reinstatement of Lukoil's earlier disputed oil contracts with Saddam Hussein. Such an opening is unlikely to benefit only American corporate oil.

Critics argue that the American empire has never had a consistent, settled strategy. Its allies and opponents have been selected according to what is convenient at the time. The goal of American imperialism at any one time, it is said, has been to sort the "preferred" barbarians from the "unwanted", and to convert the former (these days, those offering secure access to oil) into vassals. The division of barbarian states helps keep down the costs of imperialism by enabling some states to be written off. The consequence might be local wars and proverbial basket cases. The UN could in principle provide a limited management system for recalcitrant colonies of modest functional value (as in Bosnia, Somalia and East Timor, for example). And, just as servants may eat the crumbs from the table of the masters, even defective states might fulfil some functional role by providing export oil and energy opportunities.

This model of subjugation is said to describe American policy in Central Asia since the region opened up to Western oil following the demise of the Soviet empire. American governments containing many strategists and politicians with a background in oil (Dick Cheney, Brent Scowcroft, James Baker, Zbigniew Brzezinski) were hardly likely to be slow to envision the significance of this petroliferous region. The ultimate prize, according to some, was to have the region policed by American imperialism while Russia turned its back in order to focus on its political and energy relationships with Europe. Local wars would be doused and pliant states expected to toe the imperial oil line. The "barbarians at the gate" would thus be kept apart and made dependent. This is not, of course, the patchwork outcome that has materialised. The region is far from one of settled US hegemony. Both Russia and China have become increasingly active in the region in relation to oil, gas and energy. Many state players hang on the edges seeking targets of opportunity.

Some have also argued that the American empire is increasingly vulnerable to "blowback" (to use Chalmers Johnson's term).[59] That is, it could face the damaging consequences of overstretching. The vision is of an America too deeply engaged on too many fronts across the world, fighting one war too many, and becoming entrapped in barbarian counter-strategies. A number of locales for such blowback have been suggested, such as Pakistan, Saudi Arabia or Egypt. Others could be Iran and North Korea. It is argued that the de facto American empire is already overextended and that the cost (however defined) merely of maintaining the status quo is set to rise inexorably. Certainly the full-scale costs of the Iraqi adventure have been enormous and may become a burden.

As America has been drawn into a global war on terror – perhaps not so much by design as by default – it has had to build its architecture of empire on the hoof rather than with strategic purpose and wisdom. Corporate oil is left to adjust to the new circumstances. Now the search for Al-Qaeda and related networks extends to over 60 disparate states. In this venture America faces rising challenges and the risk of unintended consequences. Some have even termed the war on terror the "Third World War". It is questionable for how long such a venture might be sustained.

Is this, then, one of those ventures – not unparalleled in history – that could lead to long-term imperial demise? Nothing, after all, lasts for ever in the mould of empires. Even the Cold War lasted less than half a century, at cost to the Soviet empire. Some, however, argue that the American empire is exceptional and may follow a unique course. Its heartland alone occupies a space that is greater than that of many entire empires in the past. It is bounded by vast oceans that once were thought to provide untouchability. Beyond its heartland the empire still has a truly global reach. Moreover, America could by some estimates be home to 500m people by 2050 (more than the combined population of the enlarged EU). The American version of empire may stand to make the earlier Roman model look distinctly provincial.

The doctrine of selective regime change could be taken as one method to secure the oil supplies that the empire needs, but it may not be the only or even the ultimate strategy. America's need for control of key oil resource zones might be accomplished with multiple mechanisms ensuring a mix of client states, partners, quasi-colonies and other subservient polities. Any such strategy, however, will incur opposition from the barbarians and costs to the empire. In time the American electorate may prefer some more minimalist version of imperialism or even a reversion to a republic in fact and not just in name.

What of American corporate oil in this context? At present its foundations remain steady. It is true that it faces more competitors. Some of these players, however, are also essential partners. And some may even fall into American hands through acquisition. Yet the industry faces considerable challenges in the developing world. As examined in Part 3, companies now need somehow to meet more demanding ethical and political scruples of liberal democracy while maintaining a hard-edged competitiveness in operations within barbarian lands where very different kinds of regimes rule. Widespread resource nationalism also reduces its options for growth. At the same time, monopolies of best practice once commanded have receded and American corporate oil is no longer the only source of capital or expertise available. The barbarians have increased choice in allies and compatible players.

In the circumstances, therefore, an unchanging corporate oil world based solely on American modes and political edicts now seems unimaginable. Here, however, we should note that industry and state are not always as closely aligned as critiques of the "military–industrial–oil" complex so readily assert. Some naive commentators, for example, attribute the discovery and development of oil in the Gulf of Guinea directly to the American government's wish to diversify away from the Middle East – as if it clicked its fingers and, lo and behold, the oil was found and produced. In fact, though the fast-growth development of the Gulf of Guinea today coincides with America's strategic aims, it has resulted from companies such as ExxonMobil and Chevron having been in place for some decades rather than from any government edict. It is also an arena of intense competition in which state players are more prominent than before.

Indeed, the interests of the American government and industry often diverge – not only over sanctions but also, notably, over crude oil prices. While the government may favour cheap oil and energy, corporate oil is less disconcerted by higher prices. Price rises can be good news for corporate cash flows. The American empire and corporate oil, therefore, do not fit together like hand and glove. They follow different and sometimes independent trajectories.

Powerful unilateralism is neither the only, nor necessarily the best, strategy available. Indeed, there is evidence of a more differentiated policy – one increasingly involving foreign partners. Not only have America's military operations involved alliances with other countries: it is likely that the task of developing Iraqi oilfields will be an international one. American corporate oil would probably not, on its own, be able or willing to bear the entire portfolio capital and risk burden entailed by this mammoth task. The likelihood is that companies from Europe, France, Russia, China, Asia, Latin America, the Middle East and elsewhere will also choose to be involved.

An American empire with preferences for partners and allies is a far from unlikely prospect. In 2006 Senator Richard Lugar introduced an Energy Diplomacy and Security Bill to the US Senate. This sought to establish strategic energy partnerships and hemispheric energy forums with allied countries. The US–UK Energy Dialogue (2003) looks set to lead to joint energy co-operation in a worldwide but selective strategy. Following meetings in Crawford, Texas, and London, this task-force recommended that the US–UK alliance should draw together their foreign and energy policy objectives on promoting security and diversity of oil and gas supply, integrating international energy investments within host countries to meet social and development challenges, upgrading clean energy technologies and somehow expanding US–UK energy trade.

A closer liaison between fragmented Anglo-American empires of oil was the aim. Most notable in this démarche has been the search for international partners worldwide. The parties seek a co-ordinated strategy to enhance Russian and Central Asian oil and gas development, look for mitigation measures for Middle East oil dependence, encourage large consumers such as China and India to play a role in oil emergency preparedness, and enhance the oil climate in Sub-Saharan Africa for foreign oil investments.

The US would like to reduce its dependence on foreign, especially Arab, oil. In the meantime, if only slowly, a new world oil order is gradually being forged as new Kings of Energy (LNG, oil sands, GTL, biofuels) have been sought to partially replace, in effect, Saudi Arabia and a number of Middle East hotspots. Russian petro-tsars have already been courted (though with limited success), African oil chiefs solicited and ex-communist Central Asian *nomenklatura* entreated. Mexican and Colombian state elites form part of this global realignment and the Brazilian oil handmaiden has been promised marriage by 100 suitors. Most of the 60 oil-producer countries worldwide have attracted some American and/or Western investment interest. The hoped-for opening up of Iraq and de-sanctioning of oil states such as Sudan (even eventually Iran) would extend this strategy of global access. All of this may with time serve to considerably reduce in part America's heavy reliance on a few desert oil monarchies.

Corporate oil, meanwhile, has been reluctant to abandon its long-harboured ambitions to penetrate Middle Eastern oilfields. Its resistance to long-term exclusion is based not simply on inbred stubbornness or corporate nostalgia but also on a long-cherished dream of commercial manna from the oil heavens.

It may be within the soft underbelly of empire – that is, inside the barbarian lands – that the greatest long-term risks are lurking.

This is a quasi-invisible world, with a multitude of players and many hostile forces, where hegemony may not guarantee success. It forms the sea in which corporate oil must now swim or sink, amid sharks and predators, where its enemies find their natural environment. It may not be classic warfare that turns out to be the final arbiter of access in the many power struggles in play. Many different strategies exist with which to counter corporate oil's assumed superiority. Somalia was a demonstration of the extreme risks, and perhaps a model for many more environs to come. But more conventional mechanisms – asset confiscation, renational-isation, oil tax imposts, forced contract migration – have all been deployed and may become more common.

In the non-OECD world, praetorian regimes with or without warlords are more common than often thought. They generate their own dynamic and are difficult to reconcile with the dictates of the oil investor market and the suave executives who inhabit corporate oil's echelons. Anti-globalisation and anti-capitalism movements might even turn out to be the softer side of this new equation of hostility.

This "other world" is increasingly becoming integrated into the cartographic matrix of portfolio choice for corporate oil. All this could thwart the advent of one global oil village in which could take place some sort of rational global oil pillage. Much of this new reality is beyond the control of any empire of oil. Hence corporate oil has sought to find replacement survival mechanisms other than just the long arm of the American government or sound advice from Western capitals to offset partially the potential need to exit some oil zones.

At the same time, it is by no means certain that the American public will entertain the long-term logic of an imperial future with its associated commitments and costs. The humiliating failure in Somalia, desire to exit from Iraq and end involvement in Afghanistan, resistance to the onerous tasks of long-term nation-building, apparent decline in altruism abroad, domestic backlash to real and perceived international hostility, fear of overstretch and lack of an established territorial tradition of imperialism – all these may militate against extending the strategy at some time in future.

Corporate oil, more so than in earlier times, may well find itself, therefore, left more on its own in the more widely contested world of the future. Its efforts to "do good while doing business", through social responsibility investments, suggests a partial engagement, at least while assets remain in situ. Much of this endeavour is, however, primarily related to project economics and portfolio life cycle: it is designed to show the softer side of invest-ment wisely using a limited budget. It can never be an open-ended attempt at fundamental social reform with permanent engagement aimed at transformation. Such an approach should thus be seen for what it is: one of strategic and tactical opportunism aimed at defending core oil interests and protecting brand image within barbarian lands on behalf of the corporate oil empire.

Corporate oil portfolio must of necessity be kept "risked", dynamic and adaptable, and it must churn, carrying only "so much" managed political risk. If protective security is needed, it is likely to be negotiated, co-opted, or imported, sometimes only temporarily. A more flexible, self-sufficient approach of this kind might provide greater stability than would reliance on distant empire. The promotion of American values and culture overseas may not in any case require in-place governors or proconsuls. Equally, corporate oil does not need an office in every country in the portfolio, although many large ventures require this. Temporary task-forces often prove rather more effective.

Nevertheless, support for social projects around oil persists in high places. Some want to save the world from itself, others to save themselves from the world. Corporate oil would like both, so long as the bottom line remains black. Civil wars and some conflicts now last longer than before, creating a highly dangerous environment in the regions that oil companies must inhabit out of necessity. Excess disorder has the unhappy tendency to breed more, in a cycle of recrimination and revenge. Thus if turbulence is more than likely to mark the future, even corporate benevolence will prove inadequate to the tasks of stabilisation.

Corporate oil typically seeks to stand aloof in most local disputes in the hope that neutrality will protect continuity, a task fraught with difficulty in Nigeria and Colombia in the past. This may be a case of hope over experience if increasing numbers of people in the oil lands experience more desperate living conditions. Where sheer survival is at stake, the virtuous few can hardly expect their proclaimed neutrality to be accepted. The logical outcome is for them to become targets, whether for the aggrieved or some organised hostile entity. There are in many communities ingrained mythologies about oil (with "images" of oil richness and endless bounty) that companies find hard to displace or "right-size". For many, especially the uninformed and misinformed, corporate oil may seem like a fount of largesse that can solve all problems. In the barbarian oil world, corporate oil can therefore be readily depicted as an ideal scapegoat for social ills and bad times.

Demographic projections suggest that the population of low-income countries will rise by over 2 billion within the next quarter of a century. Human development indices indicate that this will add to a number of stresses. Much demographic expansion is forecast for key oil states – Saudi Arabia, Yemen, Pakistan, Indonesia, Nigeria and elsewhere in Africa, including the Maghreb (notably Algeria and Egypt). Many fast-growth populaces are found in Muslim societies. Many experience flawed governance and varying degrees of authoritarianism, led by elites and power structures that will in time face tests of continuity and even regime legitimacy. The existence of anti-Western extremism in such places does not augur well for an easy entry and problem-free sustained presence on the part of corporate oil.

Past attempts by rich global elites to treat backwardness (which was how poverty was once described) with mega-aid programmes have largely failed. Poverty alleviation strategy, the new mantra of the global institutions including much of corporate oil, has fared little better. Some African leaders have absorbed such grand ambitions about radical poverty reduction and high economic growth rates that a failure to meet expectations may be all but inevitable. When things "go wrong" a target for blame will be needed. Neither the American empire nor corporate oil will wish to be the scapegoat for such travesties. But whereas American political interests can be reassigned, oil wells, acreage and discoveries cannot. They can only be farmed-down, sold or abandoned.

Strategies aimed at mitigating the oil–poverty nexus have already engaged the best in corporate oil management over many decades. They look set to preoccupy them even more in future, as this strategy has become embedded even while imperfect. The companies have in effect set themselves up as veritable aid agencies – often without the knowledge and experience of the latter. Why they should succeed where others have failed is not clear and is seldom discussed.

The failures in managing the oil–poverty nexus have also led to new ideas in this arena. Plans for foreign oil revenue management intervention in Chad, mediated by the World Bank, look likely to be followed with schemes for Mauritania and, in time, São Tomé & Príncipe and elsewhere. These models for oil revenue management have yet to stand the full test of time. Most presidential regimes will find such models of global paternalism hard to stomach. They are likely eventually to take them as a slight on sovereignty, since they are intended to limit the opportunities for plunder that only flawed systems can properly provide. Whereas in Western countries some leaders acquire wealth before entering politics, in developing countries many politicians enter politics in order to acquire wealth. Well-intentioned Western schemes often fail to appreciate this fact and the facts of local history. Western liberal mindsets are often found to be out of sync with hard-edged political realities.

As a model, the foreign regulation of state oil revenues has had imperial precedents, as empires controlled far-flung domains.

In the 21st century it may become one of the parameters of global dissonance that will be difficult to shift towards on a worldwide scale. It implies a centralist world, one that must function with widespread political acceptance. Behind its façade will be needed the casting aside of entrenched cultural matrices, the split local/ foreign management of social and economic conditions, and reformation of deeply ingrained political mindsets.

The events of 11 September 2001 and since may have provoked a new American global activism to enhance oil interests within the US foreign policy agenda. This does not mean that the war on terror will stretch into the pre-emptive elimination of all rogue states and the reconstruction of the increasing number of failed and failing states. The task would be too colossal. Notwith-standing the evidence of mounting engagement from America – in Bosnia, Kosovo, Afghanistan, Iraq, Central Asia and elsewhere – the voices of caution can already be heard articulating the dangers of nation-building, overstretch, mission creep and the need for the selective targeting of resources. From some attuned to the future, the motto may even be heard: "It's the oil, stupid."

It is this argument that might ultimately carry the day: selective engagement, for strategic necessity, and oil of course. It has already underpinned many engagements and is certainly one key to Central Asian commitments. The new alliance that was sought with Russia likewise had oil as one central concern. Hence too the signals emitted from America to Nigeria to quit OPEC, lift oil production constraints on super-majors and become strategically enmeshed with the US (along with Angola) as a complementary source to Saudi Arabia and the Persian Gulf as a supply of oil.

Regime stabilisation in Baghdad could come to exceed all these initiatives in its ultimate oil implications, although this is much dependent on unrealised production trajectory outcomes. Theoretically, it could provide the US some access to the world's second-largest reserve base, while posing a related energy threat to Iran, and allow for the "Mother of All Strategic Oil Realignments": the displacement of Saudi Arabia as prime arbiter of the spigot in OPEC. This notion, however, is far away at this time.

One earlier dream has already faded: the idea that, with a resurgent oil-producing and exporting Russia, in close alliance with America, as well as Iraqi crude output of approximately 6 MMBOPD, a new dynamic could be reached in world oil markets. Such grand strategy has been overtaken by events. In a different context, a reshaped empire of oil perhaps made all the sense in the world. Had it been realised, American corporate oil – to the enragement of its competitors and enemies – would have benefited hugely in the bottom line. It may now never happen, as the world open to the classic empires of oil has in several ways been diminished.

6

Prometheus bound

Corporate oil's universe has been shrinking. We need to examine the forces that have increasingly constricted corporate oil. These include OPEC, oil sanctions and strategic moves to reduce oil dependency. Here we examine each of these in turn and consider also the consequent shape and direction taken by the contemporary oil industry.

To the list of constraints indicated we should add the strategies adopted by America's rivals as empires of oil, especially in regard to resource nationalism. The significance of national oil companies has been growing for a long time, diminishing the choices and toughening the terms available to corporate oil in the process. The growth of this industry segment is far from over, especially as national oil companies increasingly join the lunge for equity oil beyond their own borders and close more deals between states. The typically close and growing relationship between government energy and foreign policies in many countries means that further state intervention in oil may be expected.

Some national oil companies (notably CNPC and Gazprom)

are now likely potential acquirers of Western oil companies, though bids from such quarters may encounter political resistance (the opposition to CNOOC's attempt to acquire Unocal signals a case in point). Though no consumer counter-cartels have so far been formed, there have been a number of energy dialogues, with co-ordination between major importers mooted. On the upstream buy-side, India and China have already co-operated in joint asset acquisitions, with purchases made in Syria and Colombia as the first fruits of this strategy. The net effect of acquisition activity from state players has been to raise asset prices, bonuses and bid threats in the corporate oil world. Much company asset acquisition has been pre-empted.

The interrelated impacts of government policy, national oil companies and resource nationalism (see more in Chapter 7) have had the effect of forcing change in corporate oil's portfolios. The typical portfolio of the super-majors now holds larger-scale ventures in fewer countries and also reveals a shift to deepwater and unconventional oil, upgraded gas/LNG and GTL assets. The state dimension within the global upstream has not yet reached a zenith. The final destination is unknown but perhaps inevitable.

✳

Corporate oil stands in a paradoxical position with regard to its sometime nemesis, OPEC. On the one hand, higher prices assist margins and profits; on the other, lower prices stimulate demand and maintain oil as the fuel of choice. Though segments of the industry seem dependent on the idea that OPEC controls price (it cannot do so wholly) and cannot see beyond it, an OPEC-free world would not necessarily be to corporate oil's detriment. OPEC has, however, been skilled in maintaining the myth of its omnipotence and the necessity for managed and so-called stable prices. In the 21st century so far, this has occurred or been managed in only an upward direction.

Given the importance of OPEC to world crude supply, specific political risk considerations in these countries could affect world markets in the future. This would especially be the case under conditions of low oil prices and competitive challenges in energy supply, with growing inter-fuel competition from gas, and the impacts of new technologies on tar sands and heavy oils (in Canada and Venezuela). There are also security and strategic risks on the horizon. There will almost certainly be a number of political transitions inside key OPEC states in the coming years, and many constitutional arrangements in place for non-democratic states are not reassuring.

Ultimately, corporate oil would wish to gain full access to reserves in key OPEC territories such as Saudi Arabia, Kuwait, Iran and Iraq. In the current dispensation most such opportunities are currently off limits despite limited gas openings in Saudi Arabia and perhaps restricted potential for North Field access in Kuwait.

A number of important oil-exporting states remain outside OPEC but have on occasion, more or less willingly, co-operated with it. These have included Russia, Mexico, Norway and Angola (before 2007). Few appear keen to join the cartel (Angola was an exception), and some almost certainly never will. Oil diplomacy, however, requires at least some pretence of co-operation. In the case of gas, even now co-operation is taking shape with on/off discussions on a world gas cartel.

With oil openings in Central Asia, West Africa, Russia and elsewhere, as well as plans by OPEC members such as Nigeria, Libya, Algeria and Venezuela to expand production beyond their OPEC quota, this club is likely to remain an imperfect price-managing entity. The spread of royalty-based models with non-proprietary high production growth at low cost – now inserted into Algerian, Venezuelan and Nigerian contracts – stands to undermine precision in the political control of output by OPEC. This would be especially true if an oil glut were to emerge, since this would place OPEC states in conflict with their investor interests and create an OPEC world bifurcated between state and private oil production models.

In some ways, OPEC is an institutional relic of the Cold War era. Yet with a history stretching over nearly half a century, it is one of the great survivors in the oil landscape. There could emerge longer-term threats to its survival. They include inter-fuel competition from unconventional oils, gas/LNG, condensates, nuclear, biofuels and coal. Members such as Gabon and Ecuador left OPEC when their interests did not coincide with production austerity, despite membership cost burdens being cited as the reason at the time. Ecuador has been seduced into returning to the fold. Nigeria and Angola (and even perhaps Algeria) could in time confront a similar dilemma about continued membership. OPEC's leadership – occasionally contested, but with Saudi Arabia dominant (having the UAE, Qatar and Kuwait on its coat-tails most of the time) – is coming under threat for political, social and security reasons. It remains to be seen whether, if ever the Saudi oligarchy changes, OPEC can retain its cohesion.

Even with the global trend towards greater diversity in potential reserve dispositions and oil export supplies, as well as a worldwide decline in old-fashioned geopolitical models and institutions, the odds appear to be on continuity for OPEC. It is a landmark that corporate oil will either want or need to continue to live with – though its survival does not guarantee that its influence will remain at the same strength in the very long term.

*

Another limitation to corporate oil's access to global oil reserves is the sanctioning of states. Sanctioned oil states currently include Iran, Sudan, Syria, Cuba, Myanmar and Somalia. As well as sanctioned states, there are also those, such as North Korea, to which oil-related restrictions apply. Iran is a sanctioned state that is an OPEC member with huge oil and gas reserves. Others are new, promising producers/exporters. A third tier consists of states with smaller and largely untested oil potential. Common aspects of sanctions include restrictions on upstream investments and energy financing, and constraints on exports and related regional impacts on pipelines.

The implications of sanctions are various. Significant, new, sizeable, high-potential exploration, development and production ventures in both oil and gas may be curtailed for some years. World supply of crude oil is probably reduced and higher prices may result. OPEC is enabled to allocate quotas without full recognition of the potential entitlements of Iraq and Iran, so facilitating quota-enhancement, as well as easing OPEC's task in sustaining higher oil-price levels. The development of the oil industry in countries such as Sudan and Syria has been retarded because they have been able to attract only limited sources of investment for exploration and a narrow range of companies into their plays.

Players once or now in place in countries such as Iran, Syria and Sudan may have gained a privileged position by not having to face the entry and competitive thrust of American corporate oil. Advantaged players in such regimes have at times included Total, Gazprom, Petronas, CNPC, ENI, BHP Billiton, Lundin Oil, OMV, Shell, Slavneft, Talisman and PetroCanada. There are also companies that have sought rights in Iran such as Japex and Inpex that in the end lost out on the Azadegan field development.

Disadvantaged players have included, most notably, Conoco (once interested in Iran), American corporate oil in general and companies with interests under *force majeure* (such as, at one time, Hess, Conoco, Marathon and Occidental in Libya). Exxon-Mobil was once keen on entry to Iran but has been unable to proceed because of sanctions. Several companies have been interested in specific states (notably Talisman and Chevron in Sudan). A number of independents would under normal conditions have targeted Iraq, Iran, Libya, Sudan and possibly Syria. Some national oil companies from sanctioned states have been weakened by the restrictions – notably INOC (at the time), NIOC, NOC-Libya, Sudapet (Sudan), Syrian Petroleum Company and MOGE (Myanmar).

The division between in-place and excluded companies has distorted the global competitive landscape. It has led to a discordant set of international policies between the US and others (notably the EU, Malaysia and Russia). States such as Syria, Turkey, Jordan and Turkmenistan which lie close to sanctioned states have also had their oil business distorted. This has affected oil flows, transit revenues, gas pipelines and government revenues.

Sanctions have complicated many deals that might yet be struck for entry into Iraq. Hence a new round of global positioning will be heralded in the future. Many companies are known to have had discussions on "agreements" that could not be executed, prior to regime change. These included Total, Petrovietnam, CNPC, Lukoil, Pertamina and several Canadian companies. Many more now seek entry, and some independents have already entered Kurdistan where open solicitation has long been practised.

Sanctioned oil targets may fail to attract international financing and be vulnerable to US regulatory retaliation. Political risk insurance may be unobtainable or prohibitively costly. However, some governments have allowed or even encouraged investments in sanctioned states. France was willing to protect Total's interests in Iran in the face of US actions and potential litigation. The reasoning is that if opportunities are not taken while sanctions apply, they may not be attainable once the sanctions are lifted – as intense competition in de-sanctioned Libya has illustrated.

Overall, sanctions have had a major impact on corporate oil and the world upstream industry. At the minimum, a world of zero sanctions would have improved the global opportunity set and supply options. In consequence, some countries with lower-quality prospects would have encountered disposals or withdrawals or experienced greater difficulty in attracting new investments.

Corporate oil's risk management is complicated by the fact that much about the future of sanctions is uncertain. Corporate oil has a vested interest in seeing sanctions reduced if not eliminated. It can argue that this would facilitate commercial and geopolitical strategies to diversify assets and production. For the American government, however, sanctioning constitutes a global management tool enabling it to achieve an extra-territorial reach. It is, therefore, unlikely to dispense with sanctioning too readily. There is currently some form of sanctioning in place against a plethora of states. Furthermore, the threat from NGOs of litigation against oil companies over complicity with unsavoury regimes imposes, in effect, a layer of unpredictable private restraints that act as quasi-sanctions.

Paradoxically, sanctions have created a de facto supply control tool rather like OPEC – only under US–UN auspices. The net result is reserve and production lockout for an unspecified time, albeit with less co-ordination than might be available to OPEC. At the time that Iraq was sanctioned, approximately 15% of world reserves and 10–15% of oil production was shut in as a result, and in effect most remains so because of the continuing Iraqi conflict.

✳

At home, corporate oil faces commercial and policy challenges from industries and lobbies seeking to replace oil with biofuels, hydrogen, hybrids, coal, nuclear energy and a range of non-fossil forms of energy. Although some oil companies (such as BP and Shell) have small parts of portfolio within these domains, oil and gas remain their core game. In America, the new energy litany has combined with moves to increase security by reducing dependency on foreign supplies of oil.

As the International Energy Agency (IEA) has emphasised, oil dependency is an OECD-wide dilemma.[60] The dependency remains even though, across the world, the amount of oil needed to generate a unit of GDP has broadly halved over the past 35 years. The lack of radical shift away from oil, say the pundits, makes the West vulnerable to OPEC, Saudi Arabia and unstable Gulf producers. Though more could be done to improve America's fuel efficiency, this will at best be a long-term process.

Even some OPEC nations (Saudi Arabia, Iran, Nigeria and Algeria are expressing interest in nuclear options with the International Atomic Energy Agency) have joined the long march towards nuclear energy. This may have less to do with domestic energy needs than with the desire to free more oil for export and, in some cases, to provide nuclear defence capability. In the case of Iran, its claim that nuclear development is required by its energy needs appears somewhat flawed, given its vast underexploited gas reserves. But its nuclear interest is long-dated and Iran holds uranium reserves, while its future oil profile is under some self-induced pressure complicated by its sanctions status.

Some public statements from senior oil executives seem more likely to have given succour to competing fuel lobbies than to corporate oil. Frank Ingriselli, when president of Texaco Technology Ventures, once opined that "we're moving inexorably towards hydrogen energy ... and those who don't pursue it ... will rue it", while the former Arco CEO (then BPAmoco senior executive), Mike Bowlin, once announced that "we're in the last days of the Age of Oil". In this climate it is not surprising that politicians have spotted opportunities to mandate targets for renewables. Governor George Pataki, for example, has instructed New York State's government to meet 30% of its energy needs from renewables by 2010.

✳

Corporate oil is caught in a vice: it is locked out of many high-reserve countries, but it needs to secure large oil reserves for the future. Super-major reserve holdings are not vast compared with the OPEC states and the largest national oil companies. The question for them is how to stay in the game in the long term. Most have large cupboards of resources yet to commercialise and so there is no direct threat. Increased resource nationalism, however, is shifting the horizons of potential difficulty closer.

The rise of resource nationalism may hold risks for oil producers and corporate oil alike. This has been evident in different forms and degrees, for example in Venezuela, Kuwait, Dubai, Yemen, Nigeria, Ecuador, Algeria, Bolivia, Kazakhstan, Libya and Russia (where it has had an impact on Shell's interests, among others). This development has probably shifted a larger share of the world oil patrimony (over 90% of reserves are in state control) into less efficient hands, as national oil companies sometimes lack the necessary technologies for maximal exploitation and their efficiency levels may be lower than those of private operators in most facets of the upstream value chain. This shift may have increased global production costs. Overall, the phenomenon of resource nationalism has thwarted or slowed deal flow with private players, accentuated interstate contracts, threatened some oil production and applied a brake on exploration in selected environs – all of which by default encourages competing fuels.

Corporate oil clearly faces a struggle to stay ahead in an ultra-competitive but narrowed world. Yet stories of "the end of Big Oil", "the demise of corporate oil" and so on are premature. The industry remains a hugely dynamic, risk-taking, element in global hydrocarbons – one that many state players would like to emulate. Its responses to the kinds of resource constraints encountered include new upstream ideas, advanced technologies, improved extraction models, wider frontier plays, exploitation of immature basins, a push into ultra-deep waters, extreme exploration, digital oilfield applications, secondary and tertiary recovery techniques, upgraded global gas/LNG and GTL plants, the search for new opportunities in a widening world of unconventional oils, as well as partnerships with compatible state players.

Let us take a closer look at the morphology of the corporate oil food chain. Using market capitalisation as a criterion, there is a "Big Five" inside corporate oil: ExxonMobil, BP, Shell, Chevron and Total. They have long histories and command enormous resources of capital, technology and operating experience. ENI and COP form a second tier, with Repsol-YPF not far behind. Then there are around 15–20 super-independents with diverse Western origins (American, British, Australian and Canadian), followed by over 100 independents that span the globe in portfolio and together play a critical role. Next are found the plentiful regional players and, lastly, the minnows – the latter a fast-growing category containing around 500 companies. Rome's legions were professional and many: so too are those of corporate oil in the Western world.

Yet many merger and acquisition targets that existed prior to the current configuration of the super-majors have long gone (for example, Amoco, Arco, Mobil, UTP, Texaco, Unocal, Fina and Elf). ENI acquired British Borneo and Lasmo. Conoco and Phillips merged. Statoil has now merged with Hydro. Even so, the history of oil shows that the corporate food chain does not stay unchanged for long. There remains plenty of scope for mergers and acquisitions. Most notably, there could in future be mega-mergers, say between BP and Shell. Some merchant bankers estimate that such a move would add £93 billion in value. Obstacles to merger exist, however, and competition Eurocrats in Brussels might well balk at such a development.

The super-independents not only form potential targets for the super-majors but are also acquirers in their own right, making acquisitions of smaller or larger fish (for example, Ocean, Kerr-McGee). There has been much recent corporate activity among independents (witness Tullow Oil's acquisitions of Energy Africa and Hardman Resources) and also among the many minnows and chongololos that have emerged in Africa, Asia, Latin America and the Middle East.[61] Several state suitors (including OIL and IOC) have sought to acquire Maurel & Prom. State companies too have become more interested in corporate acquisition (as the

CNPC/PetroKazakhstan deal illustrated). Around half the reserves that have changed hands since 2003 have gone to national oil companies. This share could even rise in future.

Yet the super-majors still have an edge in certain competitive domains. They hold large cash piles and command high-performance skill sets and technologies. Many possess undeveloped ventures stored up for the future. ExxonMobil has 75 BBLS of potential in rights held in over 30 countries. Their investments in technology (Shell appointed a group technology officer in 2006) and extraction strategy are likely to advance at a greater pace. Corporate oil is diversifying its portfolios into oil sands, gas/LNG and GTL, and in some cases (BP and Shell notably, if modestly) into green fossil fuels. Longer-term ambitions concerning oil shale ventures have been reignited in the US by Chevron, where around 1 TBLS of potential is thought ultimately to be exploitable – though commercialisation remains a long way off.

The shift to gas in corporate portfolios has also been profound, even though much of the world's gas resource is under state control. Pipeline and LNG ventures, with GTL, offer new strategic perspectives. Gas has less resource concentration, no formal cartel for now, enables corporate players to exploit technology edge and presents a vastly different political risk matrix (although Shell's Sakhalin experience with Gazprom, as with others in Russia, points to some threats similar to those of the oil game).

Nor is the world wholly bounded or at the end of exploration. There is exploration potential in many fast-opening frontier countries across the developing world. The deep offshore in many littoral states has been opening rapidly. EEZ claims will expand this exploration universe, especially around the African continent, as new offshore zones become open.

Many new oil companies have been formed. Research by Global Pacific & Partners found at least 300 companies are active in Africa alone, with growing numbers now found in Latin America (especially Brazil, Colombia and Peru) and Asia, while more private equity in upstream oil is coming out of the Middle East and Europe where several new companies have also been formed.[62] Although many of these companies are tiny and often undercapitalised, they are between them making a considerable impact in primary acreage markets within frontier worlds and basins, and in marginal field plays. In many respects they form the collective cutting edge of corporate oil, as the advanced legions of empire that will shape some of its future.

These players have been joined by many domestic corporate entities in the upstream coming out of Nigeria, the Gulf of Guinea, Egypt and North Africa, the Middle East (for example Kuwait, Oman) and Latin America (notably Brazil and Venezuela). Together with the worldwide independents, they are changing the global map for corporate oil. They face many obstacles (including those around capital adequacy, risk management, threats of takeover, or strategy error) and not all will survive – but many will and yet more companies will be formed over time. This is the new face of corporate oil: just as new armies were recruited regularly to defend Rome and its empire.

7

Emerging empires

We have examined corporate oil in relation to the American empire (Chapter 5) and generic constraints on the industry (Chapter 6). We should equally consider the industry in relation to the competitive and emerging empires of oil. We focus here on Russia (especially), China, Iran and Venezuela – though there are likewise lesser powers in the global landscape (as mini-empires or incipient empires of oil).

First, however, a word of caution is advised. Many of the developments taking place in the new paradigm have worked against Western corporate oil. They include denial of access, resource nationalism and competition from state oil companies. From a Western perspective, it is easy to assume that these form part of some coherent, unidirectional, anti-Western strategy. The truth, however, is more complex. The dynamics involve a number of factors – consequences following from the end of the Cold War, the demise of unipolarity in geopolitics, the political transition or maturation of key oil states and interactions between them, and the drivers shaping the upward trajectory of the price of crude in the past few years.

It is best to consider the emerging empires of oil from their own positions and perspectives. The focus should be on their national interests, global strategies and the rationales behind them, as well as their strengths and weaknesses, relationships with corporate oil, petroleum interfaces with the West and, indeed, their partnerships and relationships with each other. No simple pattern is found, just as in Rome each barbarian tribe presented unique issues for the management of empire.

✳

Russia is, of course, a global hydrocarbon power. It has huge reserves of oil and the world's largest gas reserves. Its oil production, already large, could even potentially rise to 11–12 MMBOPD over the next few years. The government carefully controls the hydrocarbons industry through two key state companies, Rosneft and Gazprom, and manages export routes and strategies through Transneft.

Following the demise of the Soviet Union, corporate oil saw in Russia a heaven-sent opportunity that it could not afford to miss. Russian oil was attractive both in itself and as a hedge against the risks found in the Middle East. The super-majors – Shell, ExxonMobil, BP and Total (Chevron being more cautious) – were particularly aggressive. Entry was secured on then favourable terms. Since then, notably in the past few years, Russian policy has shifted dramatically towards étatisme, with local oligarchs such as Yukos first in the firing-line, and resource nationalism. The Gazprom episode over Sakhalin-2 in 2006 (see page 140), and moving in on Shell assets and equity, cast the die. Though

TNK-BP has yet to feel the full brunt of this strategic change, it has already succumbed to higher tax pressures. BP has received threats to its licence for the Kovykta giant gas field in eastern Siberia. It must have fears about the net equity it will ultimately be allowed to hold in this venture as Gazprom and Rosneft have eyes on equity stakes. COP cut a strategic equity deal, holding up to 20% of Lukoil (a private player). Total, however, remains (with Hydro) in the Kharyaga oilfield venture and received an assurance from President Putin in late 2006 over its future. All super-majors now have concerns over the Russian play.

As the Russian bear has metamorphosed into an oil bull, the wider risks to corporate oil have become evident. The changes include an increasingly assertive government under Putin, hostility to the oil barons who acquired great oil wealth through early, flawed, privatisations, a Duma that rejects past favourable investment guidelines for production-sharing agreement terms, numerous adverse changes in taxes and environmental regulations, Russian deals with foreign state oil companies, and the manifestations of a bureaucratic state with sub-Western standards in terms of public governance and the rule of law.

For critics of corporate oil, their involvement with Russia provides a new target of great interest. Governance in Russia is imperfect, the state's democracy is considered flawed, environmental standards (though improving) are lax, some oil executives (such as the ex-Yukos CEO) have been imprisoned or exiled, there is a war in Chechnya, human rights issues abound and Russian forays into the politics of the Caucasus and Central Asia have been heavy-handed. Meanwhile, Sino-Russian pipeline co-operation presents a dual target for NGO and civil society criticism.

Gazprom was established in 1992, being carved out of the old Soviet Ministry of Gas. Managed in part by the *siloviki* (ex-security services) close to the presidency, it has a grip on 25% of all gas sent to Europe, some assets in Africa (held via Gazprombank) and the Middle East, and new holdings in Latin America. A significant $3 billion deal with Bolivia has been mooted. It remains the chief middleman in the Eurasian gas play. A strategic pact, starting in 2007, has been signed with its largest customer, ENI. This provides the prospect of joint oil assets in Africa and entry into Italian gas-power markets, as well as potential LNG worldwide. Gazprom promises a continuing relationship with StatoilHydro too. The old America–Russia energy axis has eroded some distance and Gazprom is intent on becoming a global player.

Gazprom has, then, made a series of deals excluding corporate oil's super-majors, including what was to be Shtokman LNG in the Barents Sea. The giant has a global, diversified strategy for oil, gas and energy. It has potential interest with Iran over gas, and in the past Russia has enabled Iran's nuclear energy strategy. Gazprom's strategy certainly assists Russia's foreign policy, though it is driven equally by its own commercial ambitions and objectives.

Rosneft, before its acquisition of Yukos, was also intent on a global, diversified strategy (as the author knows from first-hand experience, Global Pacific & Partners having conducted its strategy review in Africa, Latin America and the Middle East). The Yukos cherry (a handmaiden of expropriation, in the eyes of some) has, however, provided richer pickings at home and the overseas strategy has been largely suspended, for now. Rosneft holds a commanding status in Sakhalin energy with equity assets in Ventures 1, 3, 4 and 5. Its status in oil within Russia is primordial.

Rosneft (84% state-owned) has now executed a $10 billion initial public offering (IPO) in London and Moscow (the world's sixth largest IPO). Many suitors looked at this opportunity to own a piece of the Russian oil empire (including Singapore's Temasek and ONGC). Rosneft plans $20 billion in new investments to raise its production to 3 MMBOPD by 2015, thereby rivalling the giants of corporate oil. It allowed BP and Petronas each an equity stake of $1 billion, CNPC $500 million and Russian investors some 47%. It seeks a market capitalisation of $160 billion by 2011 to lift it into the global league table. A second IPO is likely.

Ideas to merge Rosneft and Gazprom have come and gone. The two are co-operating rivals mediated by the state through the government, which achieves greater leverage by keeping the two companies separate. As paws of the Russian bear, together these two companies have a powerful grip on the world oil and gas game. They provide the foundations of corporate and state oil empire.

Lukoil, a private Russian private player with 16 BBLS in reserves (more than ExxonMobil), has international portfolio and global ambitions. It is already strong in Central Asia and North Africa and is emerging fast in Latin America. State threats in 2006 to revoke licences on 19 oilfields have made its portfolio abroad even more attractive as a hedge.

In the immediate aftermath of the Cold War, there was momentum towards greater co-operation on energy between America and Russia. This has much weakened over time. The two former Cold War enemies have conflicting interests. Russian oil needs high crude prices and its budget is oil price sensitive. The US benefits most from lower oil import costs. Furthermore, the rivals disagreed over the Iraq war (and Lukoil's and Russian state company contract interests there), Iran's nuclear industry and sanctions, and a range of political concerns including US military bases inside the near abroad.

The Russian grip on Europe and Central Asia, its discussions with Algeria on gas strategy and its membership of the 15-country Gas Exporting Countries Forum (covering two-thirds of world gas reserves) have raised fears of a gas cartel. These were aired at the NATO Summit in Riga in December 2006, and since. Russia's status in the G8 makes it special among non-Western players. Some even fear that Russia will shift its resource allocation eastwards, with Chinese encouragement, to the detriment of Western markets.

Russian strategy is strongly nationalist with global overtones. As the Soviet Union it was once, along with America, the ultimate arbiter of world power. Now it seeks a new place in the sun based on energy superpower status. There are some question marks over Russia's ability to achieve this ambition. They concern the country's capacity to grow and sustain oil production and rebuild its exploration game, the need for capital and advanced technology (especially in the Arctic and remote Siberian oilfields), weaknesses in the financing of the Russian army, continuing demographic attrition, Chinese encroachment in the Russian Far East, stalled economic and institutional reforms, and Central Asian resistance to Russian dominance.

A number of factors might help Russia overcome these obstacles. Oil riches have flowed abundantly into the treasury. It is cashed-up with revenue and foreign exchange. Russia's oil technicians are skilled and have a track record dating back to when the country was clearly the world's premier oil producer (even today, if gas output is included in the mix, Russia leads the hydrocarbon world). It has no OPEC quota to shackle its strategy. It also appears more ruthless in the execution of tactics and strategies to meet its grand designs. Indeed, although the oil industry in Russia needs more capital, the government argues that Western sources are not required. It may seek to limit the role of Western corporate oil to that of subcontractors providing the necessary technical assistance.

The Gazprom lunge in late 2006 to secure 50% (plus one share) in the Sakhalin-2 equity interests of Shell and Mitsui/Mitsubishi reflects the new, muscular energy politics in play. Shell has taken a beating. Russia's move was made easier by mistakes on Shell's part, including cost overruns and state claims made about environmental derelictions. This allowed Russia to play the green card to ultimate effect.

The excision of hopefuls (Chevron, COP, Total, StatoilHydro) from the Shtokman LNG venture (based on the world's third largest gas field) was another wake-up call. This put the brake on Russia's early Atlantic LNG exports and has consolidated a new gas pipeline supply option into Europe at American expense. The

British government's thwarting of a bid to take over Centrica has not stopped Gazprom in its tracks. It will rethink its strategies and is likely to return. East European states have been forced to sign revised terms for gas deals in Hungary, Ukraine and even Belarus (once considered in special relationship with Russia). The Russian pipeline strategy in the near abroad aims to ensure continued dominance despite Western counter-initiatives. The strategy has all the signs of long-term durability and the American government is concerned.

The visit of Dick Cheney, vice-president of the US, to Central Asia in May 2006 was aimed at thwarting Russia's march into the region's rich gas domain. In particular, it was intended to induce Kazakhstan to send gas by pipeline across the Caspian via Azerbaijan to Europe. This, however, is not yet a done deal. An attempt in the 1990s to get Turkmen gas going the same way proved abortive. American writ is not universal.

Russia's energy démarche with China and East Asia is in full bloom. Ventures by Gazprom and Rosneft in Sakhalin provide huge influence on the periphery of these energy-hungry giants, with two gas lines and an oil pipeline from Siberia to China promised. China and Japan compete avidly for Russian "benevolence". CNPC has signed a protocol with Rosneft for upstream oil projects in Russia. The China card is no longer a form of diplomatic adultery, but holds its own dynamic. America is one of the principal losers in this triangular game. Symbiotic interests draw Russia and China together for now, yet may contain seeds for further dispute.

In recent times the American empire has been guilty of

misinterpreting Russia's strategy. The West has been "shocked" by elements of Russian initiative – the takeover of Yukos, for example, and ex-Yukos CEO Mikhail Khodorkovsky's incarceration in a modern gulag. This dismay did not, however, thwart Rosneft's IPO in London. Russia's strategy fits into a long tradition. Where in the past it used ideological struggle and proxy wars abroad, today it is using oil and energy strategy – aggressive state players, interstate gas deals and worldwide hydrocarbon alliances all included – to build itself an empire of oil. A decade ago this was unthinkable. Now its ramifications cannot be easily contemplated.

✳

China is seen as a key concern for the American empire, despite its reclassification by the American government from an initial strategic competitor to a "strategic partner". It is large, fast-growing and oil-hungry. It has sent out its state players (CNPC/ Petro-China, Sinopec and CNOOC) to build global hydrocarbon portfolio and source energy with equity oil and LNG for China. This is in essence an "empire of oil procurement" that seeks ownership of more and more of world oil reserves abroad.

Interstate oil deals now abound. Chinese diplomacy has been shaped in the service of this ambition with bilateral agreements and major oil and gas deals struck in Angola, Venezuela, Sudan, Russia, Iran, Kazakhstan, Turkmenistan, Indonesia, Nigeria, Morocco, Equatorial Guinea, Ecuador, Myanmar, Colombia, Syria, Kenya, Niger, Papua New Guinea and even in the West (Australia for LNG and uranium, Canada for oil sands). The strategic energy links to Russia are increasingly critical. CNPC is a shareholder in Rosneft.

China has oil security high on its agenda with billions of yuan due to be spent on crude and product storage by 2020, by which time its demand for oil will be around 11 MMBOPD and imported gas inflows are expected to reach 3.6 TCF/year. The old Chinese oilfield mainstays of Daqing, Shengli and Liahoe, which between them produce 75% of domestic crude, have reached a slow-rising plateau and total oil production is growing only modestly. By 2025 China will import 75% of all the crude it requires. Reliance on imports, especially from the Middle East, makes China vulnerable. It wants to diversify this risk and at the same time ensure anchor oil contracts are secured.

China has, therefore, gone global in its search for oil and gas/ LNG in a leapfrog strategy akin to a quantum jump. This quest has brought China's oil companies into direct competition with corporate oil. Its state oil complex now controls over 7–8 BBLS in oil reserves overseas. To this must be added more LNG deals with Australia (to carry through to 2035), others concluded in Eastern Indonesia (at BP's Tangguh, in which CNOOC owns equity) and deals by Sinopec for Iranian LNG. With oil demand rising from 7 MMBOPD in 2005 to maybe 11 MMBOPD by 2020, China needs more foreign equity crude (and gas). Hence the global competition for hydrocarbon supply will accentuate. In this game China is prepared to use all the tools of cash, diplomacy and leverage it can muster. It can also execute its strategy in a co-ordinated fashion, with state companies acting hand-in-glove with the government's strategic mandates and global intentions.

Chinese state players have made several strategic acquisitions, including Nations Energy (Canada) with assets in Kazakhstan. Sinopec bought into Colombian oil by acquiring 25% of Ominex, in a joint venture (JV) with ONGC (another 25%). Both bid successfully for PetroCanada assets in Syria. Citic, the state investment arm in Hong Kong, has joined this oil asset splurge. CNOOC's $20 billion failed attempt to take over Unocal was backed by soft state funding, but it was blocked by Chevron and the US government. Husky Energy (Hong Kong-controlled) has attracted China's eye though no deal has been consecrated. Petro-China has a war chest of around $40–50 billion with which to play. China, if it wished, could consider many companies towards the top of corporate oil as potential targets.

Chinese upstream entry strategies have sought advantage from sanctioned countries (notably Sudan and Myanmar), deployed liberal use of soft-rated China ExIm Bank funds, made high-signature bonus payments for acreage and assets (in Angola and Nigeria), cut strategic JV deals (Sinopec with Sonangol), and executed many asset and block deals tied to promised infrastructure development. Chinese upstream staff are now more prevalent in the oilfields of the world outside China, including in seismic and services. They have changed the parameters of the upstream game in several zones, leaving corporate oil at some significant disadvantage and unable to match many of the package deals offered.

Increasingly, China's strategy brings it into competition with Japan. Japan imports 80% of its energy needs (99.7% of all oil). It has a worldwide oil supply network and *shogo shosa* trading houses with global equity positions, once part-funded by JNOC (and now less so, by the state successor, Jogmec). The private Japanese oil players (Inpex-Teikoku, the new combine and "national champion" designate) have to compete for global portfolio to raise *Hinomaru* oil (that is, equity oil from Japanese producers) from 15% now to 40% by 2030. This strategy intensifies the competition with China as both compete for long-term supply contracts. The two countries compete in particular for Russian energy options, including pipeline outlets. They have clashed over claims regarding the Chunxiao offshore gas field. The dragon has risen in the east, and the old imperial power is waning in comparison.

China's strategists have nonetheless myriad global concerns in oil and gas. They include a continued need for foreign investment from corporate oil, pollution costs from coal, competitive military modernisation, jurisdictional control of the South China Sea and putative Middle East destabilisation from US initiatives. A particular focus is the relationship with America, and lack of control of the Malacca Straits (through which China's oil imports pass). Perceived military encirclement of China is a strategic worry.

China's oil strategy also involves internal issues. In the west, Xinjiang province is critical for Chinese oil and gas and for the potential gas pipeline flows from Turkmenistan. Uighur militancy has dampened foreign upstream investment there. In the eyes of NGOs, corporate oil in the Tarim Basin has already been "tainted" by association with the Chinese military crackdown there. A resurgence of Uighur activism would present difficulties to both the Chinese government and corporate oil. Discovery of major reserves in Tibet at Qiantang – estimated by Chinese geologists at (a possibly exaggerated) 28 BBOE – might attract investment from abroad. It too will attract criticism in the light of Western concern about Tibetan autonomy and China's human rights record. On such matters, as with similar Western ethical sensitivities abroad, China applies a policy of non-interference as a means of deflecting any criticism.

Notwithstanding China's stunning economic growth record in the post-Mao era, the country faces severe economic challenges. These include growth differentials between the coast and the interior, massive unemployment, and advancing desertification southwards from the Takliman and Gobi deserts. Over the next ten years an estimated 100–150 million Chinese will abandon the land and move to the cities. It is estimated that China needs to create 17 million jobs a year simply to prevent unemployment rising. Given its structural needs, in terms of economic demands and national security, China is certain to continue to pursue its strategy to grow as a major empire of oil.

China's model has also caught the attention of a neighbouring giant on the move: India. Its strategy as an emergent empire in many ways resembles China's. India too is currently dependent on Middle Eastern oil (73% of its oil imports come from that region). It is expected that by 2030 India will need to import approximately 90% of its oil.

Like China, India now seeks greater geographical diversity in its sourcing of oil. ONGC, OIL, GAIL and IOC have expanded abroad and acquired more assets worldwide. A major target has been Africa, especially Sudan and Nigeria. India has begun showing greater interest in Latin America too. Its diversification strategy includes an important place for regional gas pipelines in the west from Iran and in the east from Myanmar, a venture subject to competitive Chinese and Korean interest.

Indian ambitions have often been thwarted by competition from Chinese players (and also South Korea's KNOC), such that the JV with China for asset acquisition was adopted on the basis that "if you can't beat them, join them". In this effort, however, China is playing a hedging game and is unlikely ever to allow the Indian elephant to squash the fiery dragon.

✻

In the Middle East, the empires of oil and oil producers compete for primacy in terms of state relations and for military advantage. Today new patterns are emerging. Some 80 years after the demise of the Ottoman Empire, 50 years after colonialism and less than 20 years since the end of the Cold War, America's dominance in the Middle East is seen as having come to an end. The implications could be vast and could reshape the new century.

In his article "The New Middle East", Richard Hass expects continued American influence, but on a reduced basis.[63] He foresees regional challenges from the competitive great powers, enhanced Iranian strength, continued authoritarianism, "militia-sation" (armed informal groups) and an unsettled Iraq for many years. The Middle East is expected to be a much-troubled arena for decades to come.

Iran is certainly playing its regional card. Its influence has grown particularly in southern Iraq and inside its Shia-led government. Iran is leading a Shia renaissance against dominant Sunni interests in Saudi Arabia and the Gulf states. This regional role, combined with its embedded theocracy and support for radical Islam, has led to growing international concern.

Into the Iranian cauldron, from which corporate oil has been much-excluded, have stepped the emerging empires of oil. China's Sinopec has a $70 billion deal for LNG with Iran and a 50% stake in the super-giant Yadavaran oilfield development as well as an exploration block. The portfolio will probably grow further. CNPC has an enhanced oil recovery (EOR) deal for the Masjed Suleiman oilfield. ONGC has sought a large LNG deal and a 20% stake in Yadavaran, while IOC is in the South Pars field development. Petronas holds multiple interests in Sirri A & E fields, the Munir Block, with 30% in South Pars Phases 2 and 3 (for an integrated LNG development). Pertamina and Petrovietnam both bid for Block 4 in Iran's 2005 licensing round. More Asian state players are likely to tie up with Iran in due course and existing state players will deepen their interests.

Iran's process of slow reform has stalled under its new political leadership and is hampered by sanctions. Its theocratic government stands accused of supporting militant groups (Hizbullah and Hamas). The heartland of clerical power is Qom, home to 45,000 Islamic scholars and over 50 seminaries. Iran's supreme leader, president, majlis, speaker, top judges and half the Council of Guardians come from Qom's clerical establishment. The elite are financed by vast sums in the form of Islamic taxes and alms. The source of theocratic power is the nexus between the state and NIOC, the national oil corporation.

Iran's entrenched political class has ambitious plans for regional leadership, seeking influence not only in the Middle East but also in Central Asia and the Caucasus, and indeed to become the world's foremost Islamic state. Iran's oil – the wealth it brings and the leverage on the world market that it provides – is the key to these ambitions, which are aided by high crude prices.

In the anti-Western, anti-Enlightenment tradition associated with Ayatollah Khomeini (Imam of the Era) and maintained by the conservative clerical establishment, America is the Great Satan. Iran's plans for regional hegemony, together with its support for Hizbullah and other militants, conflict with US interests. They threaten key American allies – Israel and the Gulf sheikhdoms. Western military encirclement is watched in Iran with apprehension. Iran maintains a sizeable armoury including long-range Shahab-3 missiles and is believed by many to be developing weapons of mass destruction. Its growing military capacity is intended to enable control in the Gulf of Hormuz, a crucial choke-point in the transshipment of much world oil. Confrontation with America might be expected at some point, especially if US forces remain ensconced in Iran's back yard – that is, Iraq.

External shocks could force the pace in this Persian empire of oil. Major changes in Iran, Iraq and Saudi Arabia would recast the Middle East's political chessboard. This would have implications not only for over 70 million Iranians, but also for Sunni and Azeri minorities (including Persian-related minorities) in other states, many in the Caspian region and Azerbaijan and all of them involved in the oil game.

Iran's oil industry is ageing. Recovery rates of oil-in-place at around 24% are lower than averages for the Middle East at 32%. The country's oil reserves, however, are estimated, following recent upgrades, at 137 BBLS. So far, new finds have boosted potential rather than actual production. Large discoveries have been asserted, notably in Khuzestan. The Caspian play in Iranian waters is yet to be fully exploited.

Iran is a gas giant with over 110 TCF proven and 160 TCF probable, and 940 TCF considered as its total resource. The country has the potential to be a major gas/LNG export supplier. The huge South Pars field, an extension of Qatar's North Field, might have 280–500 TCF in resources. Iran plans major export volumes by pipeline to the Gulf and several LNG ventures for Asia.

These oil and gas reserves have attracted many players. ENI-Agip and Elf signed buyback contracts. Japex, with Shell, Tomen and Inpex, sought Azadegan in south-west Iran. BG Group entered and exited. Enterprise Oil was once in Phases 6–8 of South Pars. Norsk-Hydro, Veba, OMV and Gazprom hold deals, as does Bow Valley, BHP Billiton, GDF and Tracer. Progress in Iran's LNG game has, however, been glacial compared with that in Qatar, now the world's LNG and GTL capital.

Iran once aimed for oil production to reach 4.5 MMBOPD by 2005 and 7.3 MMBOPD by 2020, but production has fallen well behind target. In early 2007 it was only 4.2 MMBOPD. NIOC is underfunded and financially weak. Several corporate oil entities active in Iran are unwilling to take further risks by expanding their ventures directly. When they do invest, companies such as Total, Shell and StatoilHydro elect the (more expensive) equity route, partly because many banks are unwilling to finance projects.

Western concerns over the looming nuclear crisis have pushed Iran to look towards rival emerging empires of oil. Moody's withdrew its rating for Iran (with "inconsistent with US policy" cited as the rationale) and Shell offloaded the Enterprise South Pars stake that it had inherited via takeover. Cepsa has exited Iran and others may follow suit in time. For now StatoilHydro, Lukoil, Sinopec, Petronas/Edison, OMV/Repsol-YPF, Gazprom, Sheer Energy and ONGC/IOC remain in place on the major field developments. BP once stated that it would not do any deals in Iran in order not to offend the US – a view that the Iranian government is unlikely to forget. Reluctance on the part of corporate oil has created an opportunity for foreign, mainly Asian, state oil companies.

The golden scenario for corporate oil would be access to both Iraq and Iran. Given current political and military conditions, together with sanctions on Iran, it is difficult to see how this is easily achievable. The US Iran-Libya Sanctions Act (ILSA), passed in 1996, was renewed for five years in July 2001. It imposed discretionary sanctions on non-US company investments in Iran of more than $20 million annually. Iran's strategy of introducing buyback contracts as a means to attract deals has had limited success. Corporate oil lobbied hard for the ILSA to be lifted but now faces the worst of both worlds – an unpromising Iraq and a closed Iran in the near future.

In the event of a confrontation with the US, the non-US players may have to revise their positions, especially if the ILSA were re-engineered to punish deals with Iran and NIOC. To date, the ILSA has not fundamentally stopped Iran's engagement with all companies. If it were to do so, this could spark a mini oil war between the US and the rest.

In *Iran Oil*, Roger Howard has provided a comprehensive view that confirms Iran's rising energy status.[64] Crude price upside has aided the migration of several states away from American dominance, a condition that Iran has readily exploited by providing new deals for foreign state players, notably the Chinese. Iran has been emboldened and America's allies weakened. There is no unified Western sanctions policy, and America's allies and competitors (notably Japan) still rely heavily on Iranian oil. Many may soon depend further on imports of Iranian gas too.

The Iran–China energy axis, part of a wider political relationship, is blooming. Energy ties with Russia, India and Pakistani are developing fast. Even South Africa, with which Iranian links go back many years through Sasol and PetroSA ventures, has had incipient deals targeted at Iranian GTL. All this accelerates the movement of the world order from unipolarity to multipolarity in a context of competitive empires.

NIOC, however, is struggling. The government treats it as a cash cow, leaving little money for reinvestment. Iran now requires much greater upstream investment. It needs more foreign partners and technologies. Even the Chinese players have invested less in their ventures than NIOC would wish. Its oil and gas are crucial to Iran's survival.

During 2004–07 Iran was the beneficiary of a crude price bonanza. State revenues, at around $45 billion–50 billion annually, have been vast. Even so, Iran's use, and mortgaging, of oil and gas proceeds to fund social programmes, subsidies for oil products and domestic gas, state bureaucracy, security investments, and policies designed to provide stability in the face of high unemployment and a rising population puts pressure on the regime.

For these reasons, Iran favours continuing high crude prices. Sustained oil price down cycles – something that Saudi Arabia could potentially induce – would be a heavy blow. It could compromise the state and undercut the $8 billion oil contracts in the hands of quasi-state companies. Iran's gas pipeline export strategies have already been much affected by US hostility and periodic difficulties with key partners (such as India) in protracted negotiations.

Whatever its travails, however, Iranian prominence in oil and gas has grown and it has already become a regional energy power, even while Gulf states elsewhere have together grown oil and gas production faster. The relationship with Syria, including co-operation such as over upstream issues, reflects Iran's growing Levantine profile. Although the decades-long tense relationship with America, exacerbated by the contentious nuclear programme, has created difficult conditions for the Iranian oil industry, in the final analysis America's strategy has failed to restrain Iran. Indeed, Iranian tentacles are beginning to stretch much further across the world – into Venezuela's Orinoco belt of heavy oil, for example. Its petro-diplomacy now has a reach expanding beyond the Middle East.

*

Saudi Arabia is a major regional competitor of Iran and a long-established energy empire, a sort of central bank of oil, which is awash with vast oil and growing gas reserves. It remains ruled by a deeply undemocratic monarchical order under the ethos of puritanical Wahhabi Islam. The leadership assumes the role of guardian of the holy places of Medina and Mecca. The House of Saud is under pressure from Islamic fundamentalists and is a target of Al-Qaeda. There is a growing gulf between the street and the elite, which derives its wealth from the rentier state apparatus.

In oil Saudi Arabia proclaims its virtues: stability, excess capacity, strength, sizeable oil reserves and an ally of the US, though this of course guarantees it numerous enemies. In general, Saudi Arabia is a price dove within OPEC's milieu. This is a great advantage to America, although the effect is mitigated by the costs incurred in defence of its Gulf oil supply and the reserves on which it depends.

Despite its long history, there is no guarantee that the alliance between America and Saudi Arabia will continue indefinitely. Some US senators have spoken of cutting the Saudi umbilical cord; and unofficial advisers have briefed the Pentagon on "taking the Saudi out of Arabia". One briefing proposed that it would be in America's interest to break up Saudi Arabia, with secession of the oil-rich Shia-populated Eastern Province as a separate state – a way of removing the oligarchy from the oil game. All this is far from official government policy, and the latter idea seems ill-advised and far-fetched. But it does indicate that Saudi Arabia cannot take American support for granted. Some inside the country suggest that new protectors should be sought. Equally, the dynasty cannot simply assume that it will continue in untrammelled power over, say, the next quarter of a century.

This uncertainty has major implications for corporate oil, which has always found the dream of access to the vast Saudi reserves seductive. When selected companies were invited to tender for gas projects in 2003, these negotiations at one time stalling, corporate oil inevitably hoped that this might be a harbinger of opportunities for oil access in future. Since 2005 such deals have been finalised. But while access to Saudi oil reserves is still the dream, there is the possibility of a triple nightmare for corporate oil – Iranian regional dominance, a turbulent or hostile Saudi Arabia and the loss of the Iraqi oil game.

Corporate oil and national oil companies worldwide have longed for access to Iraq's huge and undeveloped oil reserves. Indeed, some small players have entered (DNO and Addax in Kurdistan, Petrel Resources in the south). Big Oil's role has been consigned to the sidelines. Its activities have included memorandum of understanding (MOU) initiatives (Hydro), field studies (Shell, BP), training deals, technical assistance (Japex) and discussions over future deals (Total, ENI). There are still a number of putative deals in the pipeline, involving companies such as Petronas, Pertamina, CNPC, ONGC, Petrom, TPAO, Sidanco, ENI, Repsol-YPF, Reliance and Perenco. But security conditions preclude any real exploration and development investment.

Access to Iraq will come one day. The cartography of both world oil and geopolitics will be redrawn when it does. But access will bring with it unresolved problems. These include not only those most obvious at present – the aftermath of insecurity, insurgency and military operations – but also a mix of industry underperformance, sabotage, contractual and legal difficulties, corruption, interregional and ministry disputes, and issues over revenue sharing. Nevertheless, Iraq has the potential to become a Goliath in world oil: it has 112 BBLS in proven oil reserves and potential estimated by some at 150–200 BBLS. In all respects the Middle East oil giants – Saudi Arabia, Iran and Iraq – could all emerge as stronger quasi-independent empires of oil.

✳

Venezuela is an oil giant in the Western hemisphere but not now entrenched in the Western mould, given the anti-West political dispensation pursued by the government. With vast oil and gas reserves, it is free from any need to import crude oil. Since the populist president Hugo Chavez was elected to power in 1999, Venezuela has become a distinctive Latin empire of oil. The country has developed its own regional policy, stoking the fires of oil nationalism in other parts of South America – notably Bolivia, Ecuador and Cuba. In contrast to developments elsewhere in the world, its own national oil company, PDVSA, has been turned inwards. Yet Venezuela welcomes investment from selected state oil companies, and seeks to replace Western corporate oil. This new empire of oil provides a challenge to old models in general and the American empire in particular. In doing so, it has diminished corporate oil's options in America's once-traditional backyard.

Venezuela has taken an unpredictable turn under President Chavez. His rule has brought social discontent (especially in the early stages), a failed coup d'état, economic distortion, political turmoil and much loss of foreign investor confidence. Despite all this, the country has become reserve richer and oil revenue fatter. Its new étatisme has made it a pacesetter in the renegotiation of international oil contracts.[65]

Venezuela's proximity to the US, its key role in crude supply and its large unexploited gas reserves should make the country a natural haven for overseas private capital. But political turbulence, tougher fiscal terms and restrictive hydrocarbons laws introduced by the Chavez regime have turned corporate oil away. Taxes have been raised and equity holding rights reduced, royalties have been raised and renationalisations have taken place in the Orinoco (affecting BP, ExxonMobil, Chevron, Total and COP).

Some foreign oil companies have reduced their Venezuelan portfolio commitment, some have exited and many have revised their assessments of the future. Though many companies remain in place on a diminished basis, Venezuela is no longer the prime destination for corporate oil it once was, should be and could become. The rules of the game have been radically upset and contract sanctity violated as the traditional upstream marketplace has been displaced by government fiat, decrees and political pressures.

Within Venezuela there has been a struggle between, on the one hand, the government, the Venezuelan National Guard and armed militia (the Bolivarian Circles), and on the other, the disenchanted middle class and old elites, including former PDVSA executives. When opposition to Chavez emerged inside PDVSA, the government intervened by appointing military officials to management positions. It then reorganised the company, reoriented strategy and introduced state-driven management practices under the tutelage of the ministry and presidency. The state's grip on oil resources has tightened considerably, and new forced contract regimes for Orinoco heavy oil deals with foreign companies in 2007 continued this earlier trend.

Prior to Chavez, Venezuela was on track to raise its oil production to meet a target of 6 MMBOPD by 2007. It even contemplated leaving OPEC. Chavez changed course, supporting OPEC strongly. Implied threats to the US were damaging to new investor perceptions. PDVSA experienced turmoil, working through five chief executives in as many years – one Chavez appointee had no direct oil industry experience. Confidence has fallen as projects and production stalled.

Over time, PDVSA has lost around 1 MMBOPD in productive oil capacity. The company has taken on huge debts to finance massive dividend payments to government which have funded expanding social programmes directed at the president's political base. Venezuela's oil cash cow has been squeezed and milked. With a relatively diminished export capacity, the government has become increasingly tied to an OPEC-led policy of tighter market supply and control. The oil market has saved Chavez through the surge in crude prices, providing much-needed cash.

The government has benefited from rising crude prices and also contributed to them. The new Hydrocarbons Law requires PDVSA to hold a minimum 51% of all projects, and state policy has pushed this threshold higher. This reversed the trend of opening access and liberalisation that had attracted $20 billion in investments and pushed capacity to 4 MMBOPD. Higher royalty rates and tougher terms have been imposed, displacing some investors. For now the current production figures are disputed, with the government claiming production at 3.1 MMBOPD. Most analysts suggest it is around 2.45–2.75 MMBOPD, an estimate endorsed by the IEA.

The turmoil within PDVSA has led to demoralisation, oil worker strikes and a purge of 20,000 management and technical staff. All this has damaged PDVSA and Venezuela's economy (oil accounting for 80% of its export revenues), as well as its reputation in the US and elsewhere. Chavez stands accused by the US of "Cubanising" Venezuela. It remains, however, an important supplier to the US with 13% of the oil import market, even while it has looked increasingly to China as an outlet for future crude flows.

In 2004 the minister of energy and mines, Rafael Ramirez, took over the reigns of PDVSA in order to become a sort of "energy tsar" with the aim of managing this "state within a state". New policies included stripping assets from foreign companies (notably Total and ENI), added tax impositions and the solicitation of foreign state oil companies. Some Chavista factions have even called for wholesale corporate oil expulsion.

Several Big Oil players have remained in Venezuela. They include COP, ExxonMobil, Total, BP, PetroCanada and Repsol-YPF. The main new entrants, however, are state oil companies, drawn from the new empires of oil. They include Gazprom, CNPC, Petropars (Iran), ONGC, StatoilHydro and Petrobras.

Venezuela has 80 BBLS in conventional crude reserves and

235 BBLS in heavy oil. Intevep (PDVSA's research and development arm) estimates Orinoco resource potential at 1.3 TBLS. If extraction rates were to rise from their current rate of 9% to, say, 30%, accessible reserve estimates would rise by 160 BBLS. These are not reserves that anyone can wholly ignore. Chavez knows it and corporate oil has had to bend to the Bolivarian winds of change.

A regional and global démarche by Chavez in oil and energy has accompanied this revolution against corporate oil. For China, Venezuela provides a hedge against Iran and the Middle East. It has upgraded its Venezuelan upstream assets and strategic oil supply links. It has signed infrastructure and crude export deals designed to increase oil flows to China in five years, reaching 1 MMBOPD by 2012.

Much is expected from state players in syncrudes projects. Venezuela hopes to expand this production by 2 MMBOPD over the next six years at a cost of $40 billion. It may well need corporate oil partners to provide heavy oil technologies. The government's overall strategy is for production from conventional and heavy oil to total 5.8 MMBOPD by 2012. In gas development, a new hydrocarbon leg is being built in Venezuela, ONGC is intent on entering the Plataforma Deltana and an LNG plant, long delayed, is now scheduled. Brazil has signed 15 energy co-operation deals with Venezuela in the past two years. In 2005 a PDVSA-Petrobras pact was signed covering Orinoco ventures and the Mariscal Sucre LNG project. The corporate shake-up in Venezuela has been substantial and negative to Western interests.

The Bolivarian strategy has changed the oil face of the Venezuelan and Latin upstream, making state control more pronounced. Venezuelan oil initiatives to assist Uruguay, Ecuador, Bolivia, Argentina and 13 Caribbean countries form part of a design to develop regional vehicles in the form of PetroAndina, PetroSur and PetroAmerica as Latin state-dominated polar opposites to corporate oil.

The Venezuelan model is not the only one in Latin America. But it has influenced resource nationalism in Bolivia and the restructuring of YPFB, contract deals within Ecuador (whence Occidental has retreated) and the political/investment climate. In Argentina, Enarsa has been established as a new state player (with private equity holders, including provinces). Overall, the Andean zone has been turbulent of late as more state control and tougher terms have been introduced and state companies (domestic and foreign) have become more prominent. Corporate oil has partially retreated, except in Colombia and Peru.

Against the recent étatiste trend in the region, Brazil, Colombia, Peru, Trinidad, Suriname, Chile and Central American and Caribbean states stand in contradistinction. Brazil has opened up acreage for nine years now with successive ANP licensing rounds. Even though it remains state-directed, Petrobras has majority private equity and a growing global portfolio. It is required to compete in Brazil. In Mexico, however, the state still controls all oil and gas through Pemex. Overall, étatisme in some form remains a widespread Latin American phenomenon while events have squeezed the terrain and some choice opportunities for corporate oil.

✳

Africa has become the great new frontier in the upstream world. Recently its acreage map has been radically transformed. Corporate oil is well embedded in the continent and new independents, both foreign and local, have proliferated. National oil companies from around the world, notably China, India and South-East Asia, including African players such as Sonatrach and PetroSA, have expanded their asset base and can now be found in situ across Africa.

The global scramble for Africa has returned with a vengeance, this time focused on hydrocarbons.[66] Elsewhere the author has noted the continent's high oil and gas reserve potential and opportunities.[67] This offers the world's empires of oil a partial offset to Gulf oil dependence and resource nationalism elsewhere. There will be around 12 MMBOPD produced in Africa by 2010, with more to follow, keeping the continent as a growing net crude and gas/LNG exporter.

Oil development has become the key driver in many African economies. There will, however, be no unified African empire of oil just as Africa has never been, and most likely will never be, united as one entity. Nkrumah's dreams of African unity belong to the past. The Organisation of African Unity (OAU) invocations for the same failed. The African Union (AU) will probably meet a similar fate. Yet nostalgia for unity is a powerful influence on the political mindset among the continent's politicians. There will, however, be several important mini-empires of oil in Africa.

Hydrocarbon resources provide a critical basis for the African future and much will depend on how wisely they are exploited.

They offer a much stronger foundation for development than dependence-inducing aid programmes do. Michael Holman (an old friend of the author's) has noted the flaws of relying on the missionary aid that some in the West wish to push down African throats.[68] Africa has received aid worth more than \$350 billion since 1971. It has led to atrophy in the sinews of the state. Africa's mini-empires of oil at least provide locally sourced funds, even if they are not always efficiently used.

Increasingly, foreign empires of oil are competing with each other in Africa. A new scramble is afoot: for oil and resources. The French *chasse gardée* in oil has long become a thing of the past. The West has multiple oil players involved, from all the super-majors to many chongololos. US interest in African oil and gas is both high and growing. The demand for oil security helps explain the military commitment in the region.[69] An advantage of West African access, compared with other key oil regions of the world, especially the Gulf, is that there is no choke-point to be navigated to European and US markets.

China's strategy in Africa is based on a triple focus: oil, trade and investment.[70] The number of Chinese presidential visits and high-powered delegations has grown along with investment in oil exploration and, especially, production assets. Africa offers China portfolio diversity and competitive terms. Some foreign and African companies have acted in effect as Trojan horses for Chinese oil ventures (for example, in Sudan and Chad), providing farm-ins to choice acreage, while Chinese players have established partnerships with many independents (for example, Vanco Energy in Morocco and Côte d'Ivoire). Elsewhere, new state company combines have been formed (with, for instance, ONGC and KNOC). China's oil footprint looms larger and larger in Sudan and Chad. Its portfolio is now sizeable in Nigeria and Angola and is growing fast elsewhere. Crude imports from Africa to China are rising and the continent is integral to the oil sinews that underpin Chinese economic growth.

Corporate oil cannot compete directly with China's ancillary infrastructure deals, soft financing, diplomatic muscle, state-to-state transactions, arms provision, sanctioned state entries and lack of engagement with human rights considerations. China's state oil players come with political weight and diplomacy attached to them as an umbilical cord. For China, the return to Africa (following earlier Maoist forays) is no longer about the business of ideology, but rather the ideology of business, notably in oil.

Despite its own vast oil and gas resources, Russia is also increasingly active in African energy. Stroytransgaz has exploration interests in Algeria's Illizi Basin (with Rosneft). Lukoil has a North African strategy and five blocks in Egypt. Russian players have entered Libya too. Angola, where historic MPLA links linger, is a key target. There are also Russian interests in Mauritania. Gazprom is already to be found in Libya (via Odex) and is seeking wider investment in West Africa in alliance with Soco (Congo-Brazzaville is a key target). Gazprom has also had discussions with Algeria over west European gas strategy. President Putin has visited African countries, including South Africa (where energy deals were discussed). Russian ministers have rediscovered Africa's tropical charms. Although in the

Cold War the Soviet Union was deeply engaged in Africa (with military assets, a diplomatic agenda and political support) and its knowledge of the continent was significant, it did not then seek African oil and energy assets. Times have changed.

The lesser empires of oil are also strongly interested in Africa. Indian state and private players now scour the continent and conglomerates like Reliance and Videocon have followed ONGC into Sudan. ONGC has a growing portfolio in key oil domains, including Libya, Sudan and Angola. It also has a JV with Mittal Energy with upstream targets in Africa, especially Nigeria. Indian diplomats have identified nine African states for oil investment. A political charm offensive has been mounted.

Brazil's Petrobras is also well implanted in African with new offshore ventures in Angola, Nigeria, Equatorial Guinea, Mozambique and Tanzania, with new projects in Libya and Congo. Lusophone connections remain effective in the old Portuguese colonies, even though Petrobras did not succeed in the Rovuma bid round in Mozambique in 2006. The Petrobras portfolio will grow further in Africa where its net equity oil production is on the rise, and where President Lula has made state visits.

Most Middle East positions in Africa have been held by private companies (such as Al-Thani from Dubai) but new state oil interest has surfaced too. Iran has direct relations with Sasol and PetroSA. Now OGDC (Pakistan's state player) is keen on prospects in the Horn of Africa with targets in Ethiopia and Eritrea. Ras Al-Khaimah Gas Commission (UAE) has taken acreage in Tanzania, and Dubai Energy made a failed bid in late 2006 for PremierOil which would have given it several African assets. Smaller private players such as Kuwait Energy have a foothold and will target more ventures, and Dana Gas recently bought Centurion Energy with a strong portfolio of Egyptian positions.

Hugo Chavez mounted an unexpected African foray in 2006 designed to spread the Bolivarian initiative into Africa. Venezuela has had strong relationships through OPEC with Libya and Algeria, from where Sonatrach technicians assisted PDVSA during the crisis in 2002 when Chavez fired top management and staff. PDVSA has co-operated with Malabo and GEPetrol since 2003. Venezuela held a Latin-Africa Oil Summit in Nigeria in late 2006 and Chavez went to the 7th AU Summit in The Gambia in 2006 to play a socialist and anti-imperialist theme. He followed on with visits to Mali (where PDVSA now has blocks) and Benin, pledging subsidised oil along the way and urging tougher contract terms on corporate oil. The harbinger of global oil nationalism sees Africa as fertile ground for oil evangelism.

This fusion of interests between foreign empires of oil and African mini-empires is redrawing the commercial and diplomatic oil map of the continent. The infusion is most notable in Africa's Big Five oil states: Nigeria, Libya, Algeria, Angola and Sudan. Meanwhile, the space available for Western corporate oil has diminished. New empires are squeezing the old.

In Nigeria the Asian players and state companies have been big winners in bid rounds with tied deals (bids linked upstream and downstream), a feature for preferential offers, often with soft or tied funding pledges attached. For example, CNOOC's putative deal for four blocks in preference allocation in 2007 has been linked to a $2.5 billion China ExIm Bank loan. South Korea and India have played the same game. More of this will be witnessed in future, and for some African states (Angola, for example) it provides a means to replace Western aid, soften traditional influences from abroad and enable avoidance of IMF strictures.

The flood of Nigerian independents into this oil game and their spread into the Gulf of Guinea has likewise changed the African landscape. In defined zones and marginal fields, foreign players must partner on restricted terms in Nigeria, while some policies favour locals. Politics has become a key arbiter of asset and acreage allocation, at times creating a "political market", and localisation is the name of this game. Calls from the West for greater transparency and scrutiny of corruption will not alter this process.

The state's grip on Nigerian oil has, however, been imperfect. New local interests in this barbarian oil world have been threatening the old order. There are secretive groups that "bunker" (steal) corporate oil's bounty. There are also irredentist entities, such as MEND, that seek to appropriate parts of the delta and cede from Nigeria, or at least secure ownership and exclusive control rights within their ethnic-controlled zones. Some militants have even formed oil companies: one (the Niger Delta United Oil Company) was awarded a block in the Niger Delta in 2006. Then there are bureaucrats inside and around NNPC and the state oil complex, including politicians, who have divested oil revenues from the treasury.[71] These clandestine entities, politicians with militia in tow (quaintly called "militicians"), and overt hostile forces damage the integrity and asset positions of both the Nigerian state and corporate oil. They reshape the playing field and likewise erode the domains once controlled by the older empires of oil.

In the Maghreb, meanwhile, resource nationalism has made some limited advances. Algeria revised its Hydrocarbon Law and imposed windfall oil taxes in 2007. Libya opened up after de-sanctioning, in contrast to many states in Africa, with tough terms and attracted fierce bidding for blocks, a process that continues. Both have allowed foreign state company entries to grow but have executed subtle bid/award strategies to retain corporate oil as a balancing force. More foreign state players inhabit these worlds than before and links with the emerging empires of oil have expanded.

In Angola, the government and Sonangol have between them maintained a tight grip on entry, awards, terms, deal flow, project sanctioning and oil politics – mediating the strategy and relationships with the foreign empires of oil in the process. The country's Chinese connections, cemented by presidential visits, have deepened to include joint ventures. Sinopec's Angolan portfolio has grown fast and it was preferred to ONGC in the reserve-rich Block 18 when Shell withdrew from its 50% equity. It also replaced Total in Block 3/80 (renamed Block 3/5 after extensions were added) at a time when relations with France were frozen. China now obtains 25% of its oil from Africa and is the second biggest importer (behind America). It intends to grow its share and is placing huge amounts of money at Angola's disposal, allowing the latter to escape the clutches of the IMF.

After a slow start, Venezuela has now entered into a relationship with Angola. It supported Angola's entry into OPEC. An interstate company oil co-operation deal was cut by Sonangol and PDVSA in late 2006. Gazprom too is seeking a tie-up with Sonangol, as are state companies from South Africa, India, Brazil and elsewhere. StatoilHydro has long been in-place. Meanwhile, Sonangol has taken its first footsteps into the wider African upstream world (with Tullow in Gabon). It will possibly in time use its riches to further increase its foreign upstream portfolio, building on some downstream links as well based on assets in the Democratic Republic of Congo and São Tomé & Príncipe.

Threats from abroad to Angola's tight control of its oil game have been rudely rebuffed. Attempts by corporate oil or NGOs to scrutinise the inner sanctums of the Angolan oil game have proved unwelcome. Now a new law (Order No. 385/06, 23 August 2006) states that only the energy minister is authorised to provide information on the Angolan oil industry to third parties. Meanwhile, in Cabinda separatists have in the past been subjected to army action, harassment of communities and new political strategies aimed at co-option of members of the dissident FLEC (Front for the Liberation of the Enclave of Cabinda). This is a piece of the oil-rich Angolan empire that is too valuable to lose.

But this is not all that is afoot. The local oil game has changed fundamentally. It is now run more on Angolan terms. Alleged corruption and the opaque management of oil monies may have enabled political elites to flourish, and the state company is now stronger than before. Beyond this, several Angolan independents have arisen (Somoil, ACR, Falcon Oil, Wodega and Grupo Gema among them) within or outside the old MPLA–Army–Sonangol complex and joined conventional bidding consortia. Some domestic companies thus provide an incipient challenge to, but also partnership with, Western corporate oil.

Under American sanctions, Sudan has looked steadily eastwards to Asia. The producer GNPOC is a consortium of Asian state companies. Petronas is a key player, China has huge acreage and growing production, ONGC has invested more and PetroSA has now got Block 14 in northern Sudan, a vast exploration terrain. Russia is courted for arms, typically supplied by China, and may well end up with oil stakes one day. The acreage, assets, equity and production of Sudan's own state company, Sudapet, have also grown, notably its equity shares in recent acreage awards (Block 17 at 34%). Local companies (High-Tech, Heglig) are prominent and close to the presidency, and state-connected companies include Khartoum State.

There is the prospect that Sudan may eventually lose areas of the ethnically mixed, Christian/animist, oil-rich south if a promised referendum is implemented in a few years' time.[72] It seems hard to imagine that southern interests will not vote for secession. NilePet, the southern Sudan state company, is already active in this area with AIM-quoted White Nile (in which it is a 50% share-holder). The competing rebels each have their own interests in related corporate vehicles and contest southern acreage awards. Complex conflicts between Nuer (Southern Sudan Defence Force and Southern Sudan United Defence Alliance) and Dinka (SPLA/SPLM) forces shape this fractious terrain. Oil companies jockey for position. Total's acreage in the south remains at some risk. Darfur's acreage remains in play. South Africa is wooing authorities in both the north and the south. Thus much of the Sudanese oil future remains unsettled. The oil stakes are, however, certain to rise. Sanctions provide state players with an edge that does not always disappear when the sanctions are lifted. A full opening in Sudan would, however, attract foreign oil companies in droves, but even then much of this new empire of opportunity will have been acquired by non-Western interests.

All over Africa new players have emerged from the bush. African-owned companies have become an increasing force and have moved to secure rights and preferences in oil (not least in South Africa under mandated black empowerment policy). The trend for local preferences will probably accentuate across the continent. Many of the developments discerned in Africa reflect global trends. They include the rise of resource nationalism and state control, the rise and spread of national oil companies, the formation of local companies with elite involvement, and the related redrawing of political oil maps. The process is powerful, holds momentum, will continue and has impinged on corporate oil's room for manoeuvre.

In Africa at large as in the world, it has been Asian national oil companies that have led the way in acreage and asset acquisition, followed by others from Eurasia such as TPAO (positioned in North Africa and Azerbaijan and looking more widely into Latin America) and Hellenic Petroleum (ensconced in Libya and Egypt). Compared with holdings by African national oil companies, the foreign state players are ascendant, and it is Petronas, followed closely by Chinese interests – not any African player – that is dominant in this quarter.

✳

Throughout the world the oil and gas chessboard is increasingly populated with state company acreage and assets acquired in the primary and secondary transaction markets. This global lunge by national oil companies has not run its course. The pace of this development has been rapid and is accelerating. There are now over 30 national oil companies that own oil/gas positions outside their country of origin. All intend to grow acreage, assets, and equity oil and gas beyond their national boundaries. More are likely to join this competition. Many are graduating from one tier of competitor status to another – moving from backyard strategy to become regional players, then to enter the international arena on more than one continent, and ultimately to reach for global portfolio status.

This trend is not readily reversible. Some have posited that it is but a phase that will submit to a regime of low crude prices. Any down cycles in oil prices would need to be substantial and sustained over a long period to force permanent decompression in the footprint of national oil companies in the global upstream. Many states have already become enriched and would resist a return to the past. Perhaps only worldwide regime change in synchronised form in most oil states might allow such a reversal of this new paradigm. This is highly unlikely.

The emerging empires of oil have sharpened global upstream competition. They have rebalanced power and control in the world oil game. The space available to Western corporate oil has undoubtedly diminished. The new paradigm of competitive empires of oil, evident to corporate players, is one in which, returning to our analogy with the Roman Empire, Rome's oil reach has shortened and the barbarians have taken control of their own oil worlds.

This new oil world is one of ultra-competition for corporate oil and less than benign for the older empires of oil. It is a collage of diverse developing worlds in which old players must compete in radically changed conditions. It is already less stable than before and may become more so, given the character of many states and the growing threats and risks to corporate oil.

8

Barbarian oil worlds

There are more states in the world today than in any previous period. Many fall into the category of reasonable and manageable: they present the usual difficulties for corporate oil. But in others it is far less easy to do business using old formulae. There are failed or failing states (under which heading we may include dysfunctional or collapsing polities, inherently weak states, terminal states and rogue states). There are many statelets (breakaway entities, claiming autonomy but lacking official recognition, and yet controlling resources). There are also special difficulties in dealing with authoritarian or more muscular oil states – in effect, "praetorian" regimes. Large swathes of the barbarian world may be found inhabiting these various categories of complexity and risk.

The archetypal failed state is one in which the regime either dominates or vacates public space in such a way that the state becomes, at least in Western eyes, de-legitimated. It breeds instability, corruption and often violence. Power elites and mafiosi come to control these worlds, including their oil potential. The judiciary is emasculated, social entitlements are removed, investment risks rise and in some cases civil war ensues. No amount of development aid or social investment can "fix" such situations. In its dealings with failed states, the West's presumptions – based on Enlightenment notions of human rights, individualism, liberal democracy and constitutional government – often either clash or fail to connect with local lineage-related, kinship-based or communitarian traditions. Failing states often emerge from the praetorian box during a crisis and make their uneasy transition towards failure.

It is instructive to survey by geography and state-typology the most problematic areas of the world. We will consider the most pronounced threats evident today, the likely threats of the future and corporate oil's responses.

❋

Deep structural marginalisation and economic backwardness have long existed in much of Sub-Saharan Africa and even parts of the Maghreb. The African landscape has been pockmarked by multiple coups d'état (thrice recently for example, in Côte d'Ivoire, now a de facto trifurcated state), prolonged wars from Sudan to the Great Lakes, across Central Africa to West Africa, including lower-intensity armed social conflicts (notably but not exclusively in the Niger Delta). In Algeria the conflict between the state and Islamic groups has not disappeared. Zimbabwe's economic implosion with political decay has had a regional impact.

The African continent is littered with unreconstructed failed states (such as Somalia), bifurcated countries (Sudan), kleptocratic regimes (Nigeria oft cited), autocracies (based on "Big Man" syndromes), one-party states akin to feudal fiefdoms, civilian-led military governments, some imploding zones or states such as Zimbabwe and nominal or flawed "democracies". Almost all of Sub-Saharan Africa has extensive impoverished zones, including many in oil-rich countries. It is far from clear that such initiatives as the formation of the African Union and the New Partnership for Africa's Development (NEPAD) or peace initiatives in countries such as Angola, Sudan and the DRC will prove strong enough to change fundamentally the course of this problematic African history.

Although Asia has exhibited strong economic growth, it retains a number of inherited systemic risks. Myanmar is sanctioned and led by an entrenched junta. The problems of periodic instability and terrorism are managed but unresolved in Indonesia. Irredentism continued for decades in Aceh and is latent in West Papua. Terrorism has crept into southern Thailand. Despite a changing of the political guard accomplished with finesse in Malaysia, a rise in hardline Islamic parties has produced growing unease. A nuclear-armed rogue state is still in situ in North Korea. There remains in the eyes of some a question mark over the long-term compatibility of ultra-rapid economic growth and centralist-driven stability achieved so far by one-party rule in China.

There are military struggles on and off in Luzon and southern Philippines with remaining threats from Muslim rebellions, one of which, Abu Sayyaf, has attracted US military involvement in an area where the Moro Islamic Liberation Front (MILF) has been active for decades. The MILF has an estimated 12,000 militant supporters and may still represent a threat to future stability. Abu Sayyaf has been much weakened, but retains hundreds of armed supporters. The capacity of such groups to spread jihadism into South-East Asia has been a concern for America and ASEAN at large. Meanwhile, the communist-led New Peoples Army (NPA) is resurgent and considered by many to be a greater threat than the Islamic groups. The NPA, which is strongly opposed to the US presence, has an estimated 10,000 armed followers and fights on over 100 guerrilla fronts. In parts of the Philippines, it serves as the de facto government.

India is a secular and largely democratic state influenced by the Hindu nationalist Bharatiya Janata Party. But it is not immune to instabilities. The Muslim minority numbers some 150 million (12% of the total population). There has been communal conflict (notably in Gujarat). The war on terror has tended to ghettoise the Muslim community. Many have gravitated towards conservative Wahhabi-financed madrassas in a search of Islamic identity and solidarity. Kashmir remains a tinderbox notwithstanding Indo-Pakistani rapprochement. In Assam there is much discontent and separatist movements have become prevalent, such that OIL has been plagued by threats, extortion demands and disruption to oil facilities.

New uncertainty has emerged in Latin America, seen from the optics of corporate oil, as the continent has evinced a shift away from receptiveness to the global economy towards a new set of autarkic positions. In Colombia, state fragmentation has been semi-institutionalised in the form of long-term rebellions led by the FARC. These have spread into Peru and even Ecuador. Social unrest in Bolivia has had a direct impact on its core venture, Pacific LNG (now terminally compromised). The new dispensation with YPFB has yet to guarantee results for sizeable Bolivian gas reserves. Following the late-1990s collapse in Mexico, when debt conditions became aggravated, the question of whether the country should accept upstream contracts and foreign investment in oil was hotly contested but rejected, although multiple service contracts were initiated in the downstream and joint ventures cut in the gas business. Foreign ownership remains off-limits in the upstream and Pemex remains a sacred cow. In Venezuela, the political and oil orders have altered radically under the presidency of Hugo Chavez, with the "stability" achieved built on a more autarkic strategy. In Argentina, stronger economic growth is required to effect a complete reconstruction of the social order following the effects of the debt crisis, currency devaluation and a seven-year recession. In-place players such as Repsol-YPF, Total and Argentine independents (such as Pluspetrol, Tecpetrol and Perez Companc, now controlled by Petrobras) have suffered. Uruguay succumbed to an economic downturn and resurfaced with a new oil nationalism expressed in a public desire not to privatise ANCAP and the creation of Petrouruguay in the upstream. Paraguay remains a poor performer and generically weak. The Latin oil star, so prominent in the 1990s, has partly dimmed, even though its resources in oil and gas are substantial and hold great potential.

In the Middle East the (mostly well-known) sources of instability include the Israeli–Palestinian conflict, the post-invasion disorder in Iraq, the theocratic and nuclear policies of Iran,

Kurdish demands for independence, the question marks over political succession in countries such as Saudi Arabia, the growth of Islamic fundamentalism in the region, and the associated threat of cross-border and global terrorism. Few of these deep-rooted issues have quick-fix possibilities.

Central Asia is riddled with complex ancient rivalries, archaic orders, fragile and flawed democratic transitions, mixed multi-ethnic groupings, autocratic regimes and some still-fossilised states (Kyrgyzstan, Uzbekistan) caught in historical time warps. Histories of militant militaristic conflict abound. Although Russia in general provides some stability on its flanks (however inhospitable), there is further potential for the dissolution of remnants of the old Soviet empire. This is reflected in conflicts in Chechnya and Dagestan and latent potential for separatism in some Muslim-dominated republics within Russia. There is conflict between Muslims and the Chinese government in Xinjiang that spills over into Central Asia.

A number of Central Asian states particularly afflicted with the risk of instability are anchor states in the oil world. Azerbaijan, for example, has large oilfields, massive investments and an increasing Western presence. Old conflicts with Armenia and now Iran pose destabilising threats for the future. The oil industry developed fast during the rule of ex-communist President Heydar Aliyev. The country, now under dynastic control, continues to be ruled in an authoritarian fashion. (Iham Aliyev, who was first vice-president of Socar, was later groomed for the presidency.) Azerbaijan has previously engaged in war over Nagorno-Karabakh. In the US it attracts lobbies (Armenian and others) against oil investment. The contest between the empires of oil for position and ascendance in this Central Asian theatre could breed future instability.

In Europe, war in the Balkans over the dismemberment of Yugoslavia has left a legacy of ruin and an environment in which it has been difficult for foreign oil companies to operate. Even in benign Croatia there are only a few foreign players (notably Agip). Kosovo's combustible history and ethnic conflict has made conventional oil business all but impracticable. Even so, Europe represents a beacon of stability on the world stage: but its hydrocarbon resources are limited in aggregate and concentrated (much found in Norway, the Netherlands, Italy, Germany, Poland, the UK and Romania) among the 27 states of the expanded EU.

Social protests, corruption and flawed privatisations have been common in eastern Europe. Albania has experienced brutal civil conflict. In 1997 oil companies were forced to pull out of Tirana because of anarchy engulfing the capital. OMV had to make hasty evacuations, expatriates departed and PremierOil's staff had to seek UN protection. The conflict allowed local mafia (criminal clans known as *fares*) to benefit from the instability, profit from the war, and grow in and beyond the control of a (much-corrupted) police force. Following the demise of communist rule after 47 years, corruption is widespread and the legislature has become known as the "Kalashnikov Parliament". The expansion of the EU towards the frontiers of Russia, with the possible inclusion of Turkey (many years now down the track, maybe never), brings a new set of optics to the European and Eurasian stage. Russia, hovering on Europe's formal edges, is flexing its oil and gas muscles vis-à-vis western Europe and the new accession states now inside the EU from east Europe.

Wherever you look across the continents it is not difficult to find the remnants of conflict from past or faded empires, even struggles emerging from new ones in formation or ascendance. Flawed state structures, from Western standpoints, are more the norm than the exception in these barbarian worlds. They hold significant implications for the clash of oil empires to come as new balances in world oil order are sought by the contesting parties.

✵

It is obviously impossible to forecast with any precision how such systemic risks to corporate oil will manifest themselves over the next quarter of a century. By their very nature, such conditions do not follow any simple linear progression. It is clear, though, that corporate oil faces a turbulent, volatile period. It is at least possible to construct a thematic taxonomy of the main risk types, though the categories that make up this taxonomy are far from mutually exclusive.

A significant question is the risk associated with the intricacies of political succession. This applies particularly to the multitude of gerontocratic regimes that litter the political landscape, particularly in Africa, a world that at times reflects the statelets and issues found in medieval Europe. Many oil states are ruled by political dinosaurs (leaders who have been in place for decades), family dynasties or dictators with family members or cronies nominated as successors, ex-communist apparatchiks, military

or warlord leaders, theocrats or oligarchs, and some by acknowledged kleptocrats – all with question marks over the processes
of succession and their legitimacy. These moments of transition
– and there will be many over the next quarter of a century – will
prove to be testing times for corporate oil and contract sanctity,
asset protection and investments.

Many oil-rich countries still have monarchies, sultanates or
dynasties (based on hereditary families) that have been in power
for decades. These include the UAE, Saudi Arabia, Morocco,
Kuwait, Oman, Qatar and Brunei. Even when these regimes offer
current stability, there may be transitional risks when heads of
state die and their dynasties become subject to open challenge.

Several leaders that have long remained in power in authoritarian oil regimes have groomed successors from within the
presidential family, just as old European dynasties did through
royal families. Potentially, this makes for continuity, though such
transitions have a mixed record in practice and in history. Syria
has managed to accomplish such a shift. This was the mode of
transition that followed the assassination of Laurent Kabila in
the DRC and yet the country has remained unstable. Egypt and
Libya both appear as potential cases for dynastic succession for
the future. In Kenya, a variant of this family model was attempted
without success when President Moi's advocacy of a chosen one,
namely the son of former President Kenyatta, was rejected in an
election. Authoritarianism is not always unloved by corporate oil
– much oil development was accomplished under Soeharto – but
when linked to dynastic uncertainty the internal future becomes
hard to predict.

At the opposite end of the spectrum come the risks associated with very young states. Recently established states may be seen as transitional, their potential for continuity as yet unproven. They are often marked by political, religious and ethnic divisions and include groups that are armed. Such states, many of them the product of the ending of the Cold War, are numerous. They include East Timor, Sierra Leone, Liberia, Kosovo and many in Central Asia. More may emerge to add to the list.

East Timor came to independence after a long guerrilla war and a bloody transition. It inherited the Zone of Co-operation Agreement, a deal originally struck by Indonesia with Australia, with a treaty subjected to renegotiation in order to alter state oil/gas shares. This process has ended, and at the time the Greater Sunrise gas field development was pushed back towards the end of the LNG queue in Australasia, a position now partially reversed. Following UN withdrawal, new internal conflict arose; and then again before the 2007 elections a spate of disputes between Fretilin (the Revolutionary Front for an Independent East Timor) and others, with divisions sharpened between the east and west, marked the first round of open elections.

Another form of risk in uncontrolled or weak states is that associated with the concentration of private arms supplies. Some parts of the globe constitute security risks by virtue of having become repositories for vast supplies of armaments (including small arms). Examples include Yemen and the Arabian peninsula, the DRC and the Great Lakes, Colombia and the North Andean states, Somalia and the Horn of Africa, Afghanistan and Central Asia. In the Chittagong-Cox's Bazar area of Bangladesh, over

100,000 Muslim refugees (Rohingyas) dwell as exiles from Myanmar's Arakan state. Many of this group's armaments have been deployed for criminal purposes or have found their way into the hands of fundamentalist and irredentist groups. This recipe of "arms, God and rebellion", as in medieval Europe, is a volatile one, and it has made many zones of the world oilfield off-limits for normal exploration and development.

A further type of risk arises in states where the central government does not fully control its territory. In these cases, there may be regions (sometimes even unrecognised statelets) under the control of clan chiefs, rebels or mafiosi. Somalia, built on clan politics, is the classic case, but this category also includes such states as Yemen, the DRC and Colombia, as well as parts of Central Asia. In such domains there may be *force majeure*, or more commonly arrested oil development, and in some cases no exploration at all.

There are also countries in which threats to oil interests arise from a challenge to government legitimacy from home-grown fundamentalists, many hostile to the West. In Bangladesh, for example, fundamentalism has flourished. The fundamentalist Jamaat-e-Islami group, which seeks an Islamic state, holds 17 parliamentary seats. The Harkut-ul-Jihad-al-Islami organisation operates as a shadowy force, with links to banned Pakistani militant groups, and has the potential to destabilise Muslim minority relations in India and Myanmar.

In some zones the risk to corporate oil's involvement derives from conditions of more or less perpetual conflict – Gaza and Israel come to mind. The succession of failed peace treaties, intifada, suicide bombings and military conflicts indicates that Israel experiences a semi-permanent state of war. It would be a rare oil company that would be prepared to take the long view and make substantial onshore oil or gas investments in such conditions. Even offshore investments are considered high-risk. The Palestinian conflict has a regional impact involving oil-rich states such as Iraq, Syria and Saudi Arabia.

Some states form a black hole for the industry by denying open access to foreign investment as a matter of policy. In Mexico, for example, the industry remains closed to corporate oil in the upstream, despite some openings in the downstream petrochemicals industry and a few other aspects of the hydrocarbons business. Oil has long been considered "sacred" in Mexico's mythology and has been "protected" from foreign exploitation since the revolution. The state-owned Pemex remains in place as a cash cow for the government. The powerful union of oil workers remains opposed to normalisation. Even if such a reform were to occur, there would remain the risk of a reversal of policy if political conditions altered. Although there are fewer closed states than previously, much of the Middle East has denied corporate oil normal access. North Korea allows only limited access on highly constrained terms and only a few hardy small independents have been willing to enter, or allowed inside, the hermit kingdom.

There are also many states in which there are found risks from terrorism. It is to this phenomenon that we now turn as it has had a demonstrable capacity to affect world oil and Western interests, even within a few years. Some people now consider that some sort of war on terror could last for decades.

❊

Terrorism is a worldwide threat, even if it has great longevity in history, and it now has global oil industry implications. Jean-Marc Balencie and Arnaud de La Grange in their study, *Mondes Rebelles: Guerilla, Milices, Groupe Terroristes*, examine the phenomenon in over 100 countries.[73] No continent escapes this rebel world order, and none has been immune to its consequences. However, although terrorism is virtually ubiquitous, the geography of terrorism reveals certain hotspots. Of key importance are not only those states sanctioned for sponsoring terrorism but also those that in effect act as transit zones for terrorist-inclined groups. These have at times included Malaysia, the Philippines, Indonesia, Yemen and Morocco. The presence of long-term terrorism in sizeable spaces of the world (Afghanistan is just one instance) diminishes options for the exploration and oil game at large.

In addition to terrorism, corporate oil faces threats from other forms of political violence. This too is a worldwide phenomenon, with concentrations not only in the Middle East and Latin America, but also in Africa, where a bewildering array of ethnic and clan-based militia may be found. Militias of a casual or semi-organised type are also to be found in rural zones and urban slums around the world. The phenomenon of warlordism is widespread, especially in Afghanistan (par excellence), Central Asia, Yemen, Liberia, Guinea, Somalia, Colombia, Sudan and the DRC. Such groups threaten and damage corporate oil's material interests, and often deny access for exploration. The overlay of terrorism on top of these often more traditional forms of violence indicates a troubled future with no foreseeable end.

Oil's position as the life-blood of the US–Western or wider OECD economic order ensures it is a prime target for terrorism. Disruption of the oil industry, an event that has occurred several times before (Suez being one case), offers the prospect of damaging Western consumers through creating supply difficulties and accompanying price shocks. Indeed, it could provide a means of destabilising the world economy in general. The US as a dominant force in the world oil and gas industry has attracted the most attention from armed groups and the industry's enemies. In general, such antagonists focus on America's interests and by derivation American companies, though they do not necessarily restrict their attentions only to American players.

Oil is vulnerable to terrorism and hostile state actions. The cartography of the industry is less fluid than that of global terrorism. Worldwide groups such as Al-Qaeda have no inherited space to defend, no defined border for operations and no sizeable hard assets. Its cadres are highly mobile. The oil industry, for all its dynamism, moves more slowly. From *ex ante* risk assessments, through implementation, to *ex post* investment returns, projects typically exhibit long life cycles. The industry's hard assets – oil wells, installations, storage sites, tankers, pipelines and refineries – offer innumerable sitting targets. Moreover, the worldwide distribution of oil centres on a number of key crude transit choke-points, such as the Strait of Hormuz, the Suez Canal, the Panama Canal and the Malacca Straits.

Al-Qaeda has declared the destabilisation of Saudi Arabia as a key strategic aim, even if its efforts have to date been controllable.[74] It can claim success already in disrupting the once-untouched US–Saudi nexus. Global strategists in oil need always to look ahead: that the stability of Saudi Arabia and the Gulf has already been called into question is, therefore, an issue of immediate strategic concern, even though it has arisen from time to time since the 1950s. Today the threats look more profound. Further destabilisation in this volatile area – encouraged by an array of Iraqi dissidents, sectarians, insurgents and foreign jihadists – augments the impact of the initial substantive acts of terror committed. It elicits a response from the US in the form of an enlarged military footprint in the region as part of an attempt to counter terrorism in the wider Middle East.

While a worldwide jihadist network aligned against US interests is proclaimed by the antagonists, it has perhaps now grown through the leadership of Al-Qaeda. The organisation has been linked to ad hoc terrorist cells (so-called Millennium Cells, the Milani Network and the Beghal Network) and a number of regional entities in Afghanistan, Lebanon, Palestine, Jordan, Somalia, the Philippines, Indonesia and South-East Asia. Common cause has also been made with radical armed groups in Syria, Iran, Iraq, Algeria, Morocco, Chechnya, Xinjiang and Pakistan. Their proclaimed quest is often articulated as a global *jihad* aimed at radical Islamic hegemony and the establishment of new caliphates. The destabilisation created and the targeting of American and Western interests pose dilemmas for corporate oil, which is ill-equipped to deal with this putative threat. It must rely on governments and even counter-terrorism co-ordinated by various empires of oil.

✳

As corporate oil has encountered more direct threats and hostile actions in the 21st century from many sources, it has turned increasingly to corporate security to protect personnel, installations, facilities and investment strategies. Although corporate security had already become a growth business, it accelerated rapidly after 11 September 2001 and will continue to grow, especially in the fragile zones of the developing world.

The initial stage of security in oil was in analytical risk evaluation, now standard and widespread. This has been much augmented by external risk assessments designed for industry usage and executive protection. But the intellectual phase has been complemented by physical security measures. These now include contingency programmes, security audits, eco-terrorism management, alarms and camera systems, bodyguards, armed interventions, emergency plans, counter-espionage, web security, document protection, employee screening, electronic systems, debugging, hidden asset location, theft control, sniffer dogs, hostage rescue, special communications devices, surveillance, active counter-measures to identified threat, rapid-response teams, intelligence training, profiling, forensic analysis, and much more. All this has still not prevented sabotage and actions harming corporate oil's material interests.

A much-enlarged corporate security industry is now in place in the world oil patch. Over 100 companies worldwide provide these services – some global, others regional or country-specific. Companies such as Control Risks (there are many competitors) offer security advice, crisis management, investigations and risk solutions, as well as kidnap and ransom operatives. Old mercenary forces have been rebranded as private military companies, with licensing systems proposed as a way of regulating them. The headquarters and other offices of corporate oil now have more extensive security apparatus surrounding them. Executives often have armed protection when visiting sensitive countries. Many are now dissuaded from visits to difficult zones in the oil world as physical risks have risen. In some locales, staff have had to be moved out for security reasons.

Corporate oil has also had to make greater use than in the past of state security services and armed forces, especially in countries such as Colombia, Nigeria, Indonesia and Sudan. In the contemporary climate of political correctness, this brings new dilemmas as some regimes use heavy-handed methods to protect state oil interests in JVs. The tactic exposes corporate oil to complicity risks where human rights abuses provide a basis for retributive civil law actions in US courts, under the Alien Tort Claims Act, and possibly elsewhere. Such claims have been made about army involvement in Myanmar's energy game.

Threats from terrorism have also led to more intense secrecy about oil on the part of some governments and parts of the industry. The US government has, for example, removed not only nuclear power plants and transmission lines from the maps on its websites, but also hydrocarbon energy pipelines. And yet such energy maps may be found in the public domain elsewhere.

The drive to diversify portfolio will require corporate oil to deal more widely with security issues and terrorism in failed or failing states, as well as in states that do not fit such categories.

In doing so, the premiums now placed in the West on corporate respectability will be more difficult to maintain. Companies now run greater reputational risks as they are forced to make tough choices between the demands of security in pursuit of protecting operations and staff and keeping their hands clean.

This will throw up many acute dilemmas, not only for corporate oil but also for the governments commanding far-flung pieces of their empires of oil in barbarian worlds. In this arena there are sharp juxtapositions between these radically different worlds, their ethos, divergent modi operandi and the players confronting one another.

Part 3

CORPORATE SOLDIERS
AND BARBARIANS

The difficulties of operating in barbarian worlds confront corporate oil "soldiers" (boards, top executives, management, professional staff, employees, even contractors) with a number of distinct challenges. The industry will encounter these problems more often as portfolio diversification leads more companies further into riskier developing worlds.

Let us begin with the human face of the corporate players and their antithesis – the soldiers and barbarians – and cast the dilemmas in dramatic form by way of stereotypes contrasting the quintessential corporate executive with the vastly different traditional barbarians likely to be encountered. The new variant of "modern" barbarian, found largely in the West, is of a different type and needs special focus. Both act as social forces upon corporate oil and empires of oil, sometimes even in concert.

Consider first the barbarians confronting corporate oil today. What are they like? Many of them will be, in some sense, members of what we might call the "underclass". Often they will come from backgrounds lacking the types of benefits and privileges typically enjoyed by corporate oil executives. Many, not all, will probably operate on low-cost maintenance structures and with a survivalist ethic. They will know what it is like to be vulnerable and will be used to taking risks on behalf of either themselves or their social groups. They will be integrated into the local culture – its norms, history, symbolism and ethno-linguistic codes. They may well be streetwise and ruthless – some even akin to a warrior class – and a good deal more acquainted with the experience of violence than are executives from the West. Some, indeed, will be armed themselves. Confusingly, though, some will come dressed in sharp corporate suits and speak with impeccable Oxford accents.

The barbarians will not necessarily always seek juridical solutions to oil conflicts. They do not all adopt Queensberry rules. The value placed by the liberal West on the norms and ideology of political correctness is unlikely to be equally shared. Corporate oil's barbarian opponents may well envy the security and comfort of life in the expatriate compound. Many will have little to lose and much to gain.

What of the oil executives who face these barbarian opponents? They will most likely be careerists climbing the executive ladder that leads to head office. For them, status is measured by the job title on their business card. Some will be technical or professional specialists with a narrow range of vision and a habit of thinking in terms of professional norms. They will inhabit a comfortable, safety-first world of air conditioning, chauffeured cars and soirées with other expatriates, and will be used to annual vacations. Even if they learn the local language, it will usually be some version of the standard form (they will probably mostly lack a feel for the vernacular). They will often know little or nothing of the local history. Their well-trained focus on the tasks at hand and management accounts will take preference over learning to understand the "other". Underlying the whole expatriate experience is an expectation on the part of most executives that they sooner or later will be moving on to some other pasture – to be replaced of course by new, potentially gullible executives facing the same characteristic learning curves. Few will ever stay long enough to understand the myriad social complexities with which, in contrast, their barbarian counterparts are preoccupied.

These, of course, are stereotypes. But such typologies can sometimes prove quite accurate or instructive. Just try visiting the Niger Delta.

Then there are the modern barbarians, many aligned against corporate oil. Their role and impact should not be neglected in the struggle over oil. They too need space in a wide-spectrum appreciation of the conflicts found within corporate oil. Let us then examine some of these antagonists to corporate oil – modern barbarians who typically exhibit 21st-century skills and technologies, with counter-oil ideologies drawn from inside the modern world. Indeed, the old adage "He who is not my friend is my enemy, and my enemy's enemy is my friend" provides a classic definition of the "enemy" in diplomacy, politics and war. This applies equally in the oil industry.

It is useful to distinguish two types of antagonist and threat. There are the "animate" enemies – certain NGOs, politicians, governments, military regimes, militias and organised political movements – and the "inanimate" ones, namely conditions (structural, political, military, diplomatic, social and economic) that are inimical to, or hostile towards, corporate oil. They include corruption, sanctions and sometimes hostile state policies, along with circumstances such as systemic instability, periodic coups and unforeseen regime change.

These threats can cause material damage to the industry and to companies. Such costs may occur at a number of levels – notably on cash flow, top-line and bottom-line earnings, assets and their valuations, decisions over portfolio and investment positioning, share prices and shareholder value, corporate reputation or brand quality, and harm to personnel or even loss of life. Difficult and adverse conditions may likewise compromise corporate entry to certain countries, or presence and operations in particular areas, threaten the viability and lifespan of oil and gas ventures, or add to corporate liabilities in unforeseen ways.

This way of conceiving the threats to corporate oil provides us with a broader view than that of conventional risk analysis.[75] The 21st century requires a model that can relate pertinent considerations of oil diplomacy, energy politics and oil conflict to the prospects for corporate survival and success. Conventional models typically omit such factors. They can be restrictive, non-contextual and ahistorical. History and context are, however, critical in a world in which corporate oil operates within fluctuating limits to access and stability.

Standard models simply assume that conditions "hold risk" (measured or otherwise) and then apply the usual mitigation and risk avoidance options. An oil company is not, however, either a neutral or a neutered entity. It has opportunities to confront, or negotiate with, its enemies or to deploy others to do the same. The enemies are rarely anonymous (although some may not be so apparent) and may operate on a number of levels, not just those restricted to the oil game. Some antagonists have strategies that are subliminal or even hidden.

Conventional models are generally based on normative country risk assessment and presume that risks are somehow "ring-fenced" within each selected country. Here, however, we make no such assumption: it is recognised that threats may be, variously, inter-country (this risk not limited to contiguous countries), global, or related to a company's country of origin. Moreover, targets may be company-specific or even related to opportunity – as has been the case with Talisman in Sudan and BP in Tibet, for example.

Remember too that the enemies of corporate oil and the industry are not static. They have histories. Conventional models are usually content to identify, list and quantify selected threats with little regard to history or the dynamic potential embedded in the threat or the social milieu in which it is made. In assessing the enemies of corporate oil, we need to consider their origins, evolution, aims, organisation, ideology, strategy and dynamic potential. This dynamism may turn out to be the most important dimension in corporate oil's decisions over long-term investment and portfolio strategy.

Conventional country-based risk models (and even worldwide ones with regional and/or country subsets) typically underestimate enemies that are focused on issues (rather than territory) or that lack specific locations. Such enemies are, however, increasingly important in today's high-tech, globalised world. Witness, for example, the development of virtual anti-oil communities widely found on the internet.

Conventional models also fail to accommodate the role of ideology and the battle over ideas, even though many enemies to oil interests operate primarily on that level. Moreover, they often focus entirely on enemies "out there" in the external world, whereas a rounded assessment needs also to consider the "threats within" – whether they be within the company, industry, or home territories of the empires of oil.

The perspective deployed here, then, takes a more inclusive view of the enemies of oil, many of whom are of course also enemies of the empires of oil, whether traditional or modern barbarians. As we have seen, Rome could not afford to underestimate the threat from the barbarians either outside or within the empire. For corporate oil, the same is true today and for the future.

9

Targets of choice

Corporate oil is a primary focus for animus on the part of many groups across the world. As a large and wealthy industry it offers no shortage of targets. Sometimes the animus is target-specific; sometimes it is directed against corporate oil as a whole. In general, corporate oil is ill prepared to handle modern forms of such animus, despite decades of experience, and often lacks a coherent strategy to deal with it.

Although corporate oil has long been subject to hostility, the situation assumed a new complexion in the late 20th century. The development of global communications, the fragmentation of the world order, the spawning of the NGO sector, the emergence of social activism, and the application of international law all contributed to a change in the dynamics facing the world oil industry. Out of this brew has emerged the potential for international mobilisation against Big Oil and oil as an energy source. The consequent dramas are played out in various venues, often in the imperial heartland – in political parties, in the media, on the internet, at company meetings, in corridors of power, within courtrooms and on the street.

One common strategy has been to target particular companies by, for example, creating negative exposure, indirectly inflicting material damage, or exerting pressure designed to force changes in company policies, strategies, or portfolio choice. Calls for divestment from nominated zones/countries or deals are one example. The targets vary. There are some global companies that attract worldwide interest and wish not to suffer reputation risk and the associated damage to share prices; there are companies tagged, rightly or wrongly, with human rights abuses; there are those with activities related to certain countries of concern to international social movements; and there are those perceived as being against "soft issues" of symbolic importance found in left-liberal politics.

The broadest range of hostilities has been directed at Big Oil, the super-majors and companies considered to have the largest footprints – just as many hunters have concentrated their onslaughts on the "Big Five" in the African savannah (elephant, lion, buffalo, leopard and rhino).

✳

ExxonMobil has long been a *bête noire* for activists across the world. For them ExxonMobil, as the largest quoted oil company in the world, epitomises all that is wrong with corporate oil. From Aceh to Chad, and in the US itself, it has attracted sustained criticism from activists. The company may be likened to an elephant – huge, commonly seen as destructive and for many hunters the target of choice. It is also a great survivor. Yet it has some softer spots which its enemies seek to exploit, especially in spheres where consensual politics or liberal philosophy prevail and can be marketed accordingly.

Among many issues, ExxonMobil has been attacked for having taken a contrarian line on the Kyoto summit and for its coolness towards renewable energy. Critics Campaign Exxon-Mobil and the Ceres Coalition assembled a group of Christian and environmental shareholders to confront the company, arguing that its case on Kyoto rested on mis-stated science and exaggerated costs, and was designed to mislead the public. They also argued that the damage to ExxonMobil through reputation risk on such issues amounted to some $3 billion – an argument dismissed by the company. (No sound methods exist for measuring such effects.)

On environmental concerns, ExxonMobil has preached restraint. It refused to join what it regarded as a popular stampede following the Kyoto agreement, preferring to support deeper scientific understanding and performance-related engagement. The company remained a member (after Shell and BP had left) of the Global Climate Coalition, an organisation that raised doubts over claims concerning global warming and which ExxonMobil funded to the tune of $4.3 million. In terms of operations, the company's approach has been not to guarantee fail-safe systems, but rather to recognise imperfection and focus on improvements and risk management. Far from the "blue-sky world" of so-called sensitised oil, this approach is guided by cautious realism and corporate realpolitik.

The company is often attacked as a great polluter and human rights violator. Most such claims have concerned its presence in some conflict zones, and hence are based on assumed guilt by association. ExxonMobil argues that it conducts its affairs within the letter and the spirit of the UN Universal Declaration of Human Rights and it has certainly contested the terrain. It also has supported a number of human rights initiatives.

With such a barrage of criticism (coming with the territory, no doubt), ExxonMobil has not shied away from criticising activists and their arguments. Recently, especially since the appointment in early 2006 of Rex Tillerson as CEO, it has adopted a more sensitive approach. It has published an extensive review, *Corporate Citizenship in a Global World*, as a counter to past hostile press from social, environmental and political sources.[76]

Despite activist critique, ExxonMobil's performance on corporate governance has been rated by Institutional Shareholder Services at 96%, higher than all other oil companies. It also negotiated with NGOs over its activities in Chad and has published its rationale for eschewing renewable energy (why adoption of such a strategy should be "obligatory" is a mystery: it is a simple portfolio choice). In fact, after investing $500 million in renewable energy in the past, ExxonMobil decided to abandon its strategy – much to the fury of the activist world. Like others it has also detailed its social investment commitments (at $133 million each year, and rising) for all to see.

While defending its constitutional right to make political contributions in North America, it details all such allocations on its website. It ranks around 150 in the Political Action Committee rankings of contributors inside the US (about the same as BP). The company also supports the EITI process. It has, moreover, emphasised its positive impact on the world economy with annual payments made totalling $363 billion, taxes of $100 billion and so on.

ExxonMobil is no longer wholly averse to NGOs. It offers compliments to UNICEF and Africare, for example. But it is certainly selective in NGO engagements and takes trouble to identify its antagonists carefully. It backs away from single-issue and anti-business groups, as well as NGOs that are fundamentally hostile to oil development. Regarding its operation in conflict zones, it recognises that its role is that of neither a philanthropist nor a peacekeeping organisation.

ExxonMobil probably has a long way to go to soothe all its critics. There are some it will never satisfy. Over the years, it has weathered intense pressure from NGOs. Mobil (later Exxon-Mobil), for instance, came under pressure over its Arun LNG operations in Aceh. It was accused of human rights violations and collaboration with the Indonesian military, yet it had no control over the latter. ExxonMobil inherited this dilemma. The US State Department's advice to reject a lawsuit brought on behalf of Aceh parties, because it might adversely affect American interests, will have provided temporary relief, but has not necessarily ended the matter.

If ExxonMobil had been forced to contest the lawsuit, it might have suffered reputation loss. If it had lost, the company might well have had to consider a withdrawal not only from Indonesia (or Arun LNG) but also from many other countries in which there is conflict over oil. A judgement on "complicity", however specific, would have encouraged lawsuits against corporate oil around the world – from Angola and Sudan, for example, and from Colombia. One of the State Department's concerns was that the lawsuit might deter other US companies from making essential investments – something the empire of oil could not afford to risk. A further concern was that the case might provide Chinese or foreign state companies with added competitive advantage.

ExxonMobil has been challenged over numerous other operations, such as drilling in the Arctic and its involvement in the Chad–Cameroon pipeline. In the case of the latter, criticism concerned not only potential environmental impacts but also the fact that some obligatory bonus funds, due and paid under contract to the government, were later used by Chad's president to buy arms, contrary to agreement with the World Bank – a matter over which the company had no control.

Company shareholders have been targeted by various groups, such as Campaign ExxonMobil, the US Public Interest Research Group and Ozone Action. Protest against ExxonMobil has been energetic and worldwide. There is, for example, a website – www.StopEsso.org – dedicated to obstructing what it terms "the world's No.1 global warming, human rights and environmental villain". Greenpeace once depicted the Esso logo with Nazi imagery before being forced under threat of legal action to remove it from the internet because of its damage to the company brand. Several Canadian environmental groups, including the David Suzuki Foundation, Friends of the Earth, Pembina Institute, Sierra Club of Canada and Toronto Environmental Alliance, joined the Stop ExxonMobil Campaign. At the Institute of Petroleum Annual Meeting in London in 2003, ExxonMobil was targeted by StopEsso.com propaganda which accused the company of ignoring global warming, destroying our children's future, sabotaging global action and persuading the US not to endorse the Kyoto agreement. Greenpeace, which once sought to tie the company to the Iraq war, has regularly disrupted Exxon-Mobil at the pumps. In 2003 it used civil disobedience methods to block the company's UK headquarters at Leatherhead. This one protest is reported to have cost the company £1 million in direct business loss.

In late 2003 the company sought to improve its relationships with NGOs. It held a series of meetings in the UK, Belgium and Asia with various bodies including green and human rights groups. According to ExxonMobil, this strategy of engaging more openly with the outside world was not a response to the StopEsso campaign, which was not invited to these meetings. However, it would be surprising if this charm offensive were entirely unrelated to actions against the Esso brand previously undertaken by NGOs.

Attacks on ExxonMobil continue. NGOs and others blithely accused the company of a range of supposedly nefarious tactics: funding climate change sceptics; commissioning or promoting questionable scientific research findings; misusing economics; playing rich countries against poor ones; hiding behind so-called front groups (API, GCC, ICC, US Business Round Table, Global Climate Information Project and US Council on International Business); seeking to discredit reputable scientists in the complex debate over climate change; manipulating and scaring the public with the help of advertising campaigns; making barefaced, unfounded denials; and buying political influence. The list is long and growing. The company denies these allegations. They are a sign of the animus that success breeds and scale in oil attracts.

✳

If ExxonMobil may be thought of as the elephant of the corporate oil world, BP is perhaps the lion: a predator that has made many acquisitions. In the face of criticism and activism, BP has deployed a wide range of responses. These include actions on greenhouse gas reductions, a pro-Kyoto stance, devising self-applied regulations on ethics, investment in solar and renewable energy, and a range of commitments on good governance and transparency.

In particular, BP has been a pacesetter in corporate social responsibility. As Lord Browne (then CEO) once put it, it seeks to treat NGOs as co-operative partners rather than "automatic enemies". BP has established a Social Investment Business Unit. It claims to have moved from a charity model to a business performance model, decentralised along the lines of its corporate structure. At one stage BP hoped to have decentralised all its social investment within a decade. The company's distinctive approach eschews the "open chequebook" model. It links social activities to its business objectives, a practice that enlightened self-interest theorists would applaud. This work is complemented by a BP Foundation designed to channel staff efforts into charitable causes.

BP has also emphasised initiatives regarding climate change and social policy. In June 2002 it announced that it would seek to diversify its workforce by recruiting a higher proportion of women, gay, lesbian and ethnic minority workers. As part of this initiative it announced that it wanted to "lose its golf club culture" (it neglected to explain why it was sure these groups did not play golf). More than half of BP's senior management team comes from outside Anglo-Saxon culture, and 40% of annual graduate recruitment is now female. Such initiatives indicate corporate oil's responsiveness to Western liberal mores – while probably also reflecting the new realities of the graduate job market.

BP's approach, which has been very different from Exxon-Mobil's, may have had some initial success. Some critics have bought into the model, seeking to influence the company through building relationships. Yet as the number of modern barbarian groups inside Western democracies and hostile to corporate oil grows, how well protected will BP remain in future? Incidents such as the 2006 pipeline leakage in Prudhoe Bay, Alaska, and the Texas City refinery disaster revealed that BP was far from immune from flaws and extensive criticism, including from US legislators. There have been difficulties over its activities in Angola, where Global Witness's request for transparency over payments provoked the ire of the government with threats of contract cancellation for potential breaches in confidentiality.

The social investment programme has been criticised for its concentration in the West (70% in Europe and North America). Operations in stressed oil environs such as Alaska, Nigeria, Indonesia, Azerbaijan and Colombia have at times raised contentious issues. Critics point out that, although BP now invests over $100 million annually in social investments, a sum also growing, these amounts reflect only a small fraction of net profits. It is as if activists will only be satisfied if all profits go outside traditional shareholder domains.

For all the company's efforts to humanise the process of globalisation, the root cause of criticism remains: BP's core activity is founded on hydrocarbons, and the oil and gas industry acts as a magnet for relentless criticism. After all, it is from the lion's kill that vultures and hyenas feed. For many NGOs the oil industry is a boon, a basis for their raison d'être.

BP once even tried to deflect criticism by adopting a "Beyond Petroleum" campaign. It invested around $150 million in a change of logo and image. Yet the campaign came under criticism. The CEO was requested at the company's AGM to submit BP's plan for exiting hydrocarbons, a ludicrous idea. Aspects of the campaign later resurfaced – notably in corporate advertising associated with the Johannesburg Earth Summit – and are now back online. Despite some investment in renewable energy, the vast majority of BP's investment is in fossil fuels. It is after all a traditional oil company with a large gas portfolio and a growing renewable business.

A website group called SaneBP (complete with BP logo) is dedicated to reforming BP. All and sundry seem to know better how to run its business. It expresses the views of a range of investors concerned about climate change, including Greenpeace, US Public Research Interest Group and "individual socially responsible investors". The aim is to prevent all damaging oil exploration and induce a shift towards investment in renewable energy. Formed in 1999, the entity has participated in BP's AGMs ever since. In 2000, along with Trillium Asset Management Corporation, SaneBP won 13.5% of the vote on a resolution to cancel the Arctic Northstar venture. In 2001 it sponsored a resolution requesting directors to explain how BP as "Beyond Petroleum" would make the transition from fossil fuels. Although the resolution was rejected on technicalities, it was amended and resubmitted with the support of 130 shareholders (including the World Wildlife Fund and a number of local councils in the UK) holding 11 million votes, a sizeable minority.

BP's AGM acts as focus for protest both outside and inside these meetings. The annual event requires high levels of security. Protest has focused on a range of issues – claims of complicity over bloodshed in western Papua, operations in Tibet, pipeline damage in Colombia and so on. On one occasion a fake BP Annual Report was released depicting the then CEO Lord Browne as commander-in-chief of the Iraq war.

It is, however, BP's engagement in the Baku–Tbilisi–Ceyhan (BTC) oil pipeline project that attracted much criticism. A coalition of environmental and human rights groups published a review of BP's involvement, claiming it contravened the company's own stated principles. The allegations concerned poor consultation with stakeholders, the "manufacture of consent" among the villagers affected by the pipelines, the use of explicit pressures from the authorities in Turkey to assure support for the venture, and a contract with Turkey that allegedly provided for "corporate impunity" overriding the country's law. On 9 December 2003 the World Wildlife Fund took out a full-page advertisement in the *Financial Times* in London, claiming that the environmental impact of the project had not been assessed, that it failed to meet IFC, EBRD, OECD and EU standards, that it violated and harmed wetlands in Georgia, and that it damaged biodiversity. The advertisement called on banks to withhold funds until transparency had been achieved. This might have implied that it would decide when all this was achieved. In the end the pipeline proceeded and BP has answered its critics.

The lessons for corporate oil are clear and important. The aims of corporate oil and its opponents diverge. Companies such as BP are in business, in the petroleum industry, to make a profit, in response to universal demand for oil by all and sundry; whereas others are sometimes in the market to profit from the existence of the oil business. Often too concessions in one arena merely attract further critique as the goalposts constantly shift.

✹

Shell has also sought to take a leading position in the industry on social and environmental issues, though with a distinctive approach. If ExxonMobil may be likened to an elephant and BP to a lion, Shell may be thought of as a buffalo – a more sensitive, non-predatory creature, with many like-parties in the herd, which ranges far and wide while leaving a lighter footprint on the grass. Though once somewhat stubborn in character, like the buffalo, Shell too has become more transparent.

Shell's open dialogue and transparent image is one that is assiduously cultivated. This is especially so in terms of its impact on the environment, on which it publishes detailed annual reports audited by KPMG and PricewaterhouseCoopers. The company seems willing to be judged in relation to an almost bewildering range of issues: equal opportunities, human rights, political payments, regional diversity, safety, social stress and so on. This reflects a shift from the past towards the sharing and caring oil company model.

In effect, Shell has become in many ways the veritable soft side of the oil industry. It advocates open engagement, transparency and of course "sustainability", one of the new buzzwords in the reformed industry's lexicon. It has a sustainable energy programme and supported the Earth Summit and the Kyoto agreement. Yet in the final analysis, Shell must expand its hydrocarbons business. It publishes business principles, sets targets for its triple bottom line (environment, social and financial) and has its progress against them measured. The company's website invites comment and criticism – you can now "Tell Shell" whatever you like. It actively seeks dialogue with its opponents and publishes their criticisms. Its advocacy of globalisation is mixed with a message of care and compassion. Shell, it seems, is out to win our hearts and minds. It is not the only one.

Indeed, in the competition between corporate oil's public relations approaches, Shell is a market leader and has been much imitated. In many ways, Shell presents itself as the archetypal New Age oil company. It is committed to a diverse energy portfolio, including gas (which has usually enjoyed a better image than oil), solar and renewable energy, and hydrogen – the acclaimed fuel of the future. Still, and for many decades, it will remain a quintessential hydrocarbons company. Yet it needs to be recognised that all non-oil wedges in portfolio have been elected on long-term strategy and commercial grounds. As do others, Shell makes a virtue out of portfolio necessity.

As part of its commitment to evolving values, Shell has reflected some of the mantras and mandates of modern liberalism. It screens contractors for any use of forced or child labour, supports the Global Sullivan Principles, and seeks to use unarmed security (employing few armed security personnel itself). It monitors any attempts at bribery (punishing malfeasance with dismissal). Shell makes no political payments and lobbies only in line with its principles (truth, no bribes, fair competition, open dialogue, openness and transparency).

Yet although its public relations initiatives have sometimes confused and disarmed its opponents, Shell too has been far from immune from criticism and activist opposition. The debacle of its revision of reserve estimates in 2004 occasioned much criticism. The main focus for animosity, however, has been Shell's involvement in Nigeria, especially in the Niger Delta and Ogoniland, even though it has taken many steps to ameliorate conditions. The company's investment there has been vast. It now includes offshore positions in deepwater and growing LNG assets. The allegations concern complicity with military action against dissidents and armed militia in the fractious Niger Delta and claims about environmental degradation (and social costs) resulting from spillages, gas leakages, wellhead blowouts and deaths.

The history of conflict in Nigeria's delta – a long, low-intensity war involving sabotage, hostage-taking and arson, now aggravated – remains highly complex. In Ogoniland a stand-off of sorts lasted for over a decade. In 2006 Shell was forced to cede its acreage there. Several communities and ethnic groups in Nigeria are armed and active. Many are opposed to Shell's presence, though some are content to seek some sort of pay-off or to engage in "bunkering" (theft of oil). In response, Shell's Nigerian subsidiary has increased its social investment, while the parent company has fought a global public relations campaign (including major advertising on CNN). In order to comply with the Publish What You Pay campaign, Shell now publishes details of payments made to the Nigerian government, which itself has taken steps to improve its own transparency in oil matters under the Nigerian Extractive Industry Transparency Initiative (NEITI).

Although Nigeria has been the main focus for criticism, it is far from being the only one. Shell has attracted allegations in relation to a number of geopolitical issues – complicity with the Chinese government (for having investments in China and having once held shares in Sinopec) and selective activities, legal for Shell, in US-sanctioned states such as Iran and Syria, for example. It has been cited with others in a multibillion-dollar class action over its alleged support for apartheid in South Africa, a claim denied by the companies. The source of the suit was a legal party that won billions of dollars from German corporate players over matters related to the Holocaust. Corporate oil may now need to consider making some contingency for reparations for such alleged sins of the past.

With regard to the future, Shell has famously employed scenario planning. This sets it apart from many others since it has also published its views. It has used different names for various scenarios over many years, including one entitled Business Class and another called Prism, with which to explore the social future of world oil. Can these reveal insights on the empires of oil and the barbarians, and what might they reveal on strategy to deal with future turmoil if it occurs?

Business Class is globalisation writ large. The power of nation states is diminished and diffused. A new form of capitalism emerges based on regionalism and the development of mega-cities. Global elites run the world economy like a business, economic prosperity prevails, and inequality is tolerated because the disadvantaged see opportunities for improvement. The US remains as the sole superpower, though contesting power centres develop. Short-term crises occur as the world makes the transition to the model, but these are successfully managed – albeit with resentment from the periphery towards the heartland. Corporate oil emerges as a problem-solver once deal clearance engages the relevant parties, and solutions are provided and negotiated. In the new world order of Business Class, transparency grows. While governments act as referees, the consumer is king. Consumers are in a position to apply boycotts on players. Margins are razor-thin and branding becomes critical. The NGO world expands and enjoys some success in class actions. The regimes of the Middle East come under greater pressure. The great oil game continues, though players require cutting-edge strategies to compete. They focus on efficiency and the global market. Non-OPEC oil production peaks and OPEC applies price controls. Around 2010, a higher crude price of $30–40 per barrel is envisaged, precipitating a shift towards fuel-cell technology. Although oil reserves remain bountiful, there is a shift to gas. The gas game moves into overdrive. The ultimate winners are those companies with large gas positions. A lot of this mimics Shell's strategy.

The Prism Scenario ("Separate But Equal Capitalisms") is very different. Instead of the monochrome geopolitics of the Business Class scenario, Prism offers a rainbow. The interaction of culture and history creates more diverse environments. Ethnic and other differences grow in importance. Localism challenges globalism. Alternative forms of modernity emerge, including those endowed variously with Muslim and Confucian values. Global connected elites are thinner on the ground and more divided by regional patterns based on notions of identity. Agreements are less global, more local and regional. The foundations of worldwide order are shallow. Resentment over globalisation and inequality produces a slump, accompanied by calls for less dependence on the US. Fringe groups grow, some from within the middle classes, united by their opposition, though not in the form of any new consensus. The political world is fragmented between such groups as the greens, libertarians, fundamentalists and the new right. Market limitations are imposed according to hostile social agendas, the shadow side of globalisation. A long oil game emerges with focus on security and access, in which governments are interventionist. After volatility before 2010, the price of crude settles into a band of around $10–14 per barrel. Emphasis is placed on clean energy and environmental concerns. Renewables break through from around 2010 and rise to 20 MMBOE by 2020. Energy companies require a local face to operate in regional markets and "own" positions. An end-game for oil and gas emerges as oil demand slows over 2010–20. No one-size-fits-all global strategy emerges. Clearly the world has moved in different ways from either scenario, and crude prices and markets have adopted different levels, compared with when the Shell scenarios were drawn a few years ago.

Our empire of oil paradigm is more in tune with Prism. Perhaps not surprisingly, Shell avoided any scenario consisting of unmitigated contestation, comprising accentuated social fractures and even components of anarchy. Its scenarios ignore geographical and historical specificity, including the current conflict in Iraq and the possible conflict to come in Iran. They focus more on the position of elites within the global pyramid than on what those elites might do when their survival is threatened. Growing divisions between rich and poor are given little weight and are reduced to a management issue. The threats to the empires of oil are far from fully elaborated, being absorbed into new "modernities". In short, Shell's scenario planning appears compromised by its failure to give an account of the threats to the empires of oil arising from the barbarians of the modern world. It seems, moreover, to have taken too little account of rising state oil competitors, and the new wave of resource nationalism now entrenched in key oil-rich zones, as well as crude price trajectories. But Shell has been acutely aware of the growing NGO threat, worldwide social divisions and the threats that lie unattended in barbarian lands.

It seems that even now Shell is searching for a new strategy, as are others, to catch up with an oil landscape that has moved fast forward and beyond the well-known anchors of the late 20th century.

❋

All this places a premium on the fleet of foot, like the leopard. A leopard moves fast, preys all over the savannah, hunts at night and is predatory by nature. It has a high success rate in hunting. Many creatures behave cautiously in its presence. It does not change its spots. If ExxonMobil is the elephant, BP the lion and Shell the buffalo of the oil world, then Total might be the leopard.

Total has competed successfully in Africa for a long time and established a portfolio position less reliant on France and the French state. Major shareholdings are no longer French-held. The company's investments are spread widely across Asia, Latin America and the Middle East. It has long had a sanctioned state investment policy (involving Sudan, Myanmar, Libya and Iran). In these respects, Total has been smart and savvy and has rarely made serious strategic errors, although new American injunctions initiated in 2007 seek to clip its wings in Iran, and new pressures have been placed on its executives over past oil deals with Iraq under the UN Oil-for-Food Programme.

Total, like super-majors as a whole, has accorded high priority to ethics, environment, health strategy and community involvement. Following the Elf takeover, the company formed a Corporate Ethics Committee in 2001. Total now measures its ethical performance systematically. It also deals with ethics-rating agencies, belongs to the UN Global Compact, has created environmental partnerships and supports humanitarian projects. Its environmental metrics are closely measured and monitored.

Total holds extensive gas interests, including LNG, has invested some funds in renewable energy and is involved in several projects to reduce emissions. It established a sustainable development task-force. Its Fondation d'Entreprise Total promotes sustain-ability in delta and riverine zones (in Nigeria and in Indonesia). It offers explicit justification for its involvement in southern Sudan and Myanmar, arguing on the basis of constructive engagement principles. Few critics are reassured.

The NGO world, though global, concentrates in its dealings with oil more on the Anglophone world. Francophone critics of Total have focused mostly on its African operations, especially Elf's dalliances in the past with certain African leaders.[77] *L'Affaire Elf* involved a nine-year investigation into the activities of senior executives and politicians in relation to the activities of Elf and of France in Africa. This colourful case involved allegations over payments and favours allegedly given to African leaders and poli-ticians. Swiss bank accounts, French spooks, luxury outlays and mistresses (one of whom wrote a book entitled *The Whore of the Republic*) have featured in this exotic mix. Indictments and prison sentences resulted from the discovery of financial irregu-larities including a multimillion-pound slush fund, money from which was stolen by some Elf employees. Despite Total's signi-ficant efforts, some of the money remains unrecovered.

An agreement in 1997 between 35 OECD counties on anti-corruption measures helped bring this chapter to a close. Total was left with an inglorious inheritance – but stellar assets from the Elf cupboard – and the industry was tarnished with a sullied reputation, one likely to heighten the continuing attention of activists.

For all its traditional subtlety and even famous stealth, like that of the leopard, Total is likely to be subject to further scrutiny from its opponents, some of which have been known to refer to the company as Totalitarian Oil. The onslaughts must have toughened an already tough company, one that others might watch with interest.

❋

Rhinos enjoy selective habitat, do not roam the entire savannah, and have not only strength but also speed that belies their size. Before they merged with each other, Chevron and Texaco each enjoyed reputations as tough competitors, adept at survival, with a high degree of focus and less spread than comparably scaled players. Of the Big Five, we might therefore think of the merged ChevronTexaco (now Chevron) as the rhino of corporate oil's Big Five.

Chevron has developed a well-articulated, if not particularly novel, social responsibility programme called the Chevron Way. It stresses good stewardship, the company's status as partner of choice in 180 countries and a record of social initiatives on six continents over some decades. Its claim to have "touched millions of people positively" through its programmes is perhaps difficult to dispute.

Chevron has signed up to the Global Sullivan Principles and engages selectively with NGOs, notably agencies such as the World Wildlife Fund, the Nature Conservancy and the World Business Council for Sustainable Development – in effect the aristocracy of the NGO world. It spends over $75 million a year on community investments. Its main emphases are diversity, community involvement, partnerships and the environment.

It has, for example, taken action to reduce greenhouse gases, including flaring reductions at Escravos in Nigeria, gas initiatives in Indonesia and the development of cleaner fuel technologies (like GTL with Sasol in Qatar and Nigeria, for example).

The rhino may be related to its habitat. The black rhino must forage and the white rhino must graze. Chevron must now do both as it seeks to develop its portfolio, which has become increasingly weighted towards the developing world. The company has encountered action from its opponents on many occasions, including hostage taking in Escravos, asset damage in the Neutral Zone, and civil conflicts in Africa, Asia, Latin America, the Middle East and Central Asia. In the 1980s it left Sudan because of war. Now Chevron faces new threats to its pipelines in Kazakhstan and Indonesia, and it has worked with the Senior Security Policy Oversight Group in the API and upgraded its security arrangements. The rhino is a tough beast and has endured many challenges to its security. Still many hunters have it in their sights. It is unlikely to be left alone to forage unmolested, even though its global media blitz of advertising (WillYouJoinUs) since 2006 has sought to soften its image in the energy world.

✳

Even though the Big Five attracts much attention, there are many corporate animals that interest the hunters of oil from the modern barbarian world. All are fair game. The menu of issues for concern is a long one, and all putative transgressors attract the growing band of activists.

One particularly important focus for activism has been

Myanmar. Hess (then known as Amerada Hess) was forced to sell its equity stake in Premier Oil following pressure from activists, as well as the AFL-CIO's Office of Investment, over the Premier Oil's assets in Myanmar. Amerada's defence of its investment on the grounds that it was indirect proved to no avail. In Europe, the Burma Campaign UK lobbied hard against Premier. Amerada was criticised for its links: it had two directors on the Premier Board and investments in Myanmar were a key part of the Premier company portfolio, and the strategic alliance between the companies rested in part on the synergy that such investments provided. In the end Amerada sold and Premier exited Myanmar where it had been from the outset the frontier breaker, opening up plays for corporate oil – a role for which it received little thanks.

Also in relation to Myanmar, Unocal (with Total) became a target for legal claims in California. The case alleged that the company had been aiding and abetting the Yangon regime and had thereby indirectly "been responsible" for rape, forced labour and abuses suffered by peasants during the construction of a gas pipeline. This claim, denied by the companies (which had absolutely no control over government entities), was filed under the US 1792 Alien Tort Claims Act (ATCA). To qualify for an action, the defendant has only to touch down on American soil long enough to be served a summons. Even where such legal action is not successful, the plaintiffs hope that it will be enough to deter other companies. Unocal was able at the time to take heart from the court's adherence to the State Department's ruling that ExxonMobil's case in Aceh should not be heard in the US.

These cases had ramifications for many other players in the industry, including ExxonMobil and Chevron, which have faced allegations concerning human rights abuses committed in Indonesia and Nigeria respectively. Both Shell and Talisman have also been the subject of ATCA suits concerning activities in Nigeria and Sudan respectively. In the US, the "land of lawyers", where litigation has become an art form, a win for the plaintiffs in such a case could trigger a flood of class actions against corporate oil. Until such time, the uncertainty engendered was sufficient to have an impact on corporate oil's portfolio decisions. In Talisman's case, it was not senior executives driving ultimate strategy but the non-corporate world outside.

Along with Myanmar, Sudan has also been a major focus for social activism. Human Rights Watch, Pax Christi and a number of evangelical, Christian fundamentalist and Catholic groups for many years opposed the government's policies in southern Sudan related to the Dinka, Nuer, Nubian and other non-Muslim ethnic groups, including animists. These groups have actively opposed corporate oil's investment in Sudan. Christian Aid has published a critical report, *The scorched earth: oil and war in Sudan*, which increased concern in Europe. The European Coalition on Oil in Sudan has also lobbied on the issue. One outcome has been to provide enhanced entry to Asian national oil companies and reduce the Western profile in a key petroliferous opportunity for the future.

Talisman decided to exit Sudan because of pressure from the churches, NGOs and investor groups (including TIAA-CREF, one of America's largest pension funds). Talisman had a soundly articulated social programme and a framework called Sudan Operating Principles with clearly defined objectives, programme performance indicators and measurable targets. All this, however, was to no avail. Talisman endured a stock price discount of about 15–20% and had its AGMs stacked with Sudanese refugees and their minders. A residual class action remains pending. As a result of such pressure, Talisman was forced to sell one of its profitable assets. Nothing related to social conditions seems to have improved in Sudan as a result. But state competitors were advantaged. Moreover, others followed Talisman's exit. In 2003 the partly privatised OMV pulled out of blocks in Sudan following pressure from the European Parliament in alliance with an NGO coalition of 80 entities from around the world. While Talisman has acknowledged that its decision resulted from activist pressure, OMV has at times asserted that its decision was purely commercial. It may rue that choice.

A similar situation arose concerning southern Morocco offshore acreage taken by Total and Kerr-McGee. In support of claims by the Saharawi Arab Democratic Republic (SADR), activists campaigned against any oil company involvement in this zone. Both Total and Kerr-McGee (since acquired by Andarko) withdrew, the former citing technical reasons and the latter a lack of strategic fit. Though Kosmos (a farm-in partner) remained in place, largely unmolested, even small players such as Wessex Exploration (then merely considering an onshore position) experienced intense and unrelenting opposition from NGOs, especially ones from Norway, and a barrage of well-orchestrated hostile press reports.

The oil industry must also now take great care over consultancy deals and the standards placed on ethical practice. This was illustrated by the demise of the chairman, CEO and head of exploration of Statoil because of a deal made with Horton Investment for advice on field and asset acquisitions in Iran. The scandal caused a major upset in ethically pure Statoil, the then 82% state-owned flagship of the Norwegian oil industry. A secret deal with Horton worth $15.2 million over 11 years was struck in return for a number of advantages, especially entry (via Petropars, a NIOC subsidiary) into the South Pars Phases 6, 7 and 8. Horton was supposedly connected to one of NIOC's directors, related to former Iranian President Akbar Hashemi Rafsanjani, and Statoil had seemingly not disclosed its arrangement, a facilitating deal that is not uncommon. An initial payment of $5 million was to go to Horton and then be registered in a tax haven (Turks and Caicos Islands), a fact that caught the eye of internal auditors. It was alleged that this amounted to a concealed bribe, leading to police investigations and Securities and Exchange Commission enquiries. $1 billion was wiped off Statoil's market capitalisation following these disclosures, even though the company took clear steps to deal fully with the matter. Now new inquiries on earlier post-ILSA Iran gas deals in 2007 have embroiled Total and its CEO in French investigations and provoked American judicial interest.

Some lessons emerge from this case. Evidently companies such as Statoil, under majority government ownership and strong social strictures, are required to operate with one set of ethics worldwide rather than dealing with specific countries differently from the way in which business is conducted in Norway. The damage to its market capitalisation demonstrates that Statoil was unable to rely on its previously impeccable reputation. The case suggests that StatoilHydro (the new Norwegian entity) could be potentially pressured to curtail the company's activities in countries where conditions might significantly differ from the so-called Norwegian model.

Moreover, what the Total issue indicates is that the new ethics and requirements which straitjacket corporate oil can be applied retrospectively to past deals, and the targets for hostile action can be selected from any company/entity inside the corporate oil universe. Additionally, the source of actions can come from the empire's inner sanctums and engage its own state functionaries. Little wonder that all companies must now tread carefully on fragile ground – not just the Big Five and their peers in the quasi-state oil world such as StatoilHydro.

It is not only Big Oil that has attracted the attention of the political class within the bands of modern barbarians. Many other companies, from the super-independents (around 15 or so private players with market capitalisation of $20 billion–100 billion) right down to the minnows – especially American ones – are prone to activism and political interference. Conoco, before it became ConocoPhillips (COP), for example, had also been the target of such opposition. In 1995 it had to exit from a deal for the Sirri A and E fields in Iran as a result of decisions by President Bill Clinton to implement US sanctions policy. When in 2003 the company again flirted with the Iranian option, pension funds in America with shares worth $31 million forced it to desist because of America's anti-terror policy, even though the company might never have ultimately wanted to enter such a project. This incident illustrates not only the power of shareholders over directors, but also the potential for clashes of interest between the American government and corporate oil. COP has since joined the "Club of Compassion" to issue Shell-style reports on its good works.

Generally, the large players – and some smaller ones – have developed defensive strategies by boosting their social responsibility programmes. ENI, for example, which operates in 67 countries, supports the UN Global Compact and engages in health-care ventures, social projects and environmental initiatives. Also, through the Fondazione Eni Enrico Mattei, it too seeks to promote sustainable development.

The issue-specific and country-focused actions of NGOs sometimes seem nihilistic. They rarely articulate an end-game or seek to take responsibility for change *ex post facto*. Rather, the ambition appears to be to mete out punishment or seek vicarious moral satisfaction (the more important the victim, the better).

The case of national oil companies is very different from that of corporate oil. Occasionally they have been criticised by NGOs (notably PetroChina over its activities in Tibet and Sudan), but generally they have received less attention – though in some countries (notably Colombia, Indonesia, Angola, India and Venezuela) they have been attacked by armed groups. In the future they are likely to experience hostility from a wider array of opponents (with Saudi Aramco already a target for Al-Qaeda). Meanwhile, state oil players often complain about the treatment they receive in the Western media, from journalists and some politicians, and even in terms of tarnished images assumed in the West.

For the time being, however, Western firms – obliged to adapt to liberal normative order – provide softer, more tempting, targets. Those who are opposed to the empire of oil's very raison d'être and have developed a strategic agenda may prove particularly troublesome, and hence the roots of the ideas driving much opposition to corporate oil and the West deserve attention.

I O

Barbarian ideology and corporate strategy

The clash of ideas and ideologies over hydrocarbons within and without corporate oil has deep roots in a contested past. The battle lines have sharpened, and in the post-Cold War confusion among nation states, as world political schisms have realigned, both antagonists (some modern barbarians) and corporate oil have searched for new strategy.

On one side is a social composite of aligned groups hostile to oil; on the other are companies that have sought to shape responses to deflect the critics and their concerns, or at least those perceived by the corporate world as merited. Both claim to be right. For many there is no middle ground.

So first we examine some of the intellectual roots of the anti-oil movements of our times. These are diverse and disparate, and their ideas find expression in global forums, even at times within the industry. Most hold the view that the world is unjust and globalisation is designed to make it worse; that corporate oil's justifications are self-serving, with a negative impact; and that the industry should be arrested, regulated, or even put out of business. Some in the climate-change lobbies would encourage the last idea.

Then we examine how corporate oil has responded to these global challenges and how well equipped it is to deal with them. This involves understanding corporate oil's lobbies and interfaces with political elites and Western governments, both of which espouse the global order, and the key notions, notably corporate social investment and sustainability agendas, which have driven much of the thinking among the upper echelons of the corporate oil empire as a means to justify its role. It can be observed that many disjunctions exist between the ideas and strategy within the centres of oil and their critics on the peripheries.

✳

Human cosmology is rich and diverse. Its countless ideologies and mythologies include many that are inimical to corporate oil. They portray many forms of nationalism, the effects of which we have already noted – in the rise of resource nationalism and of state oil companies, the denial of access to "yanqui", "gringo", "imperialist" corporate oil, and the need for protectionism (though this appeals more to some of the weaker local or regional players). Several forms of fundamentalism (not only Islamic, but also Christian and green) also inspire virulent opposition to corporate oil.

Many ethnic, tribal or clan-based ideologies draw on mythologies that are in practice inimical to corporate oil. In Colombia, for example, U'wa mythology sees subsurface oil as the blood of the earth: its exploitation portends global catastrophe. In parts of Africa and elsewhere, the land is seen as owning the people rather than vice versa, and so rights to subsoil riches can never really be ceded by distant governments. In Australia, aboriginal mythologies attribute spiritual qualities to the land, and the Mabo land and societal claims have had an impact on oil players in Australia's onshore and even offshore.

Overriding all these local cosmologies, found in the developed world and reaching into developing worlds, exists one dominant anti-oil ideology, namely anti-globalism. As recent geopolitical events have heightened the profile of corporate oil, so the industry has become a larger target for the activities of anti-globalisation lobbies. Their groupings are many and diverse. They include (to cite a few British examples) London Rising Tide, Wombles, London Class War, Disobedience, Mayday Collective and Critical Mass. They are replicated across the Western social landscape in different guises and nomenclatures.

Not all anti-globalisation groups are fundamentally opposed to corporate oil. Some are willing to negotiate with companies, though others appear to regard dialogue (let alone compromise or co-option) as a stain on their ideological purity. Whatever the tactics or ideologies, CEOs in the oil world cannot afford to ignore their impacts or their potential resonance with a wide swathe of public opinion, especially the anti-oil groups driven by criticism of the carbonised world.

A key event in the anti-globalist calendar is the World Social Forum. This used to be held every year in Porto Allegre, Brazil, but is now hosted by a different venue each year. This boisterous event, which attracts 50,000 people, acts as a counterpoint to the

World Economic Forum, held annually at Davos, Switzerland, and designed for about 3,000 worthies drawn from the corporate, government and academic elites. Corporate oil's soldiers have typically preferred the privacy, intimacy and comfort of Davos to the rough-and-tumble of the World Social Forum and its talk shops of affiliates across the globe. This is perhaps understandable – but the result is that charges levelled against the industry at the latter go unchallenged, allowing debate to become a one-way street.

Oil is certainly not the sole focus of discussions at the World Social Forum, but it does feature on the agenda both in its own right and in relation to many other closely connected issues – notably those of biodiversity, the rights of indigenous peoples, rainforest protection, military regimes, war and human rights.[78] Indeed, corporate oil is often depicted as the human rights violator par excellence. Environmental damage, now climate change, is often taken as the direct responsibility of corporate oil: suggestions that issues of complexity apply and that multi-causal change might be involved are not greeted sympathetically.

The role and actions of government, most considered complicit with corporate oil, are also a prime focus. Many anti-globalists are conspiratorial by nature and are convinced that governments are seeking to criminalise their ideas, roles and activities – especially after their "success", in terms of attempts at direct political action, to shape the global agenda at G8 and WTO forums in Seattle, Genoa and elsewhere. As a result, governments – especially the US government and those of oil states – are treated with suspicion and are presumed to act in tandem with corporate oil at the expense of the world's poor. So close is the relationship presumed to be that corporate oil is said to "run the world".

Among the *bêtes noires* at the World Social Forum are the Washington Consensus and IMF structural adjustment and privatisation policies. These are generally viewed as instruments of global social control – the modern world's equivalent of traditional colonialism. Any deal related to the IMF or with IBRD approval is likely to be regarded, at best, as highly suspect. A particularly nefarious role is allocated to the World Bank on the grounds that it has lent over 25 times as much money to carbon-fuel projects as to so-called sustainable and/or renewable energy ventures. It has in part been recent pressure from anti-globalist movements that induced the World Bank to curtail its deals through the IFC relating to oil. That such IBRD loans have always been driven by government borrowers is simply forgotten while IFC investments are made on an equity and commercial basis.

It is widely believed among the anti-globalisation movement that the world is "up for sale" and being taken over by private equity interests. Mass mobilisation is needed to reverse this process. Global "resistance" is an abiding leitmotif. Nation states have become the captives of global capitalism, aided and abetted by the IMF, World Bank, US Treasury, WTO, UN, G7, G8, NATO and so on. Susan George, a leading protagonist of the movement and from the "old left", cites a figure of 60,000 multinationals with more than 500,000 affiliates, forming a hydra-headed system of global control.[79] The demon of Big Oil sits on the apex of this corporate collage.

Much is made of the links between the American government

and Big Oil – especially the presence of oil men such as the US president and vice-president and the contributions made to party coffers (especially the Republican Party) by oil companies. It is assumed that the link is seamless, with the government simply controlled by the industry. Corporate oil, it appears to the critics, runs the world. It must therefore be opposed.

The trend towards upsizing in the corporate oil world does not impress the anti-globalists. Corporate entities, not just oil ones, are alleged to wield more economic power than most nation states, with the top 200 multinationals said to command 25% of global GDP, yet (taking all multinationals) employing only 60 million people (the indirect job creation and multiplier impacts are ignored, of course). The oil industry, with its typically high capital intensity, is treated with particular opprobrium.

The fact that the largest oil companies each have a market capitalisation that is greater than the GDP of many states is cited frequently. Many in the public arena, even some in politics, appear to be persuaded by such judgement. One problem with this data and the implied connection, of course, is that the two figures are not comparable. Market capitalisation is a stock reflecting value. GDP is a flow like income. Normal capital/output ratios of, say, 4:1 render these supposed comparisons wildly misleading to say the least, and the comparison of income with wealth is imperfect. Moreover, it is seldom pointed out that corporate oil has multiple risks to cover, employs millions of people in the developing world and has a cost/benefit impact typically well in excess of the size of in-country portfolio (often itself substantial). Such technicalities fall on deaf ears. The debate is ideological and intended to be so.

Meanwhile, corporate oil's search for benchmarked rates of return in oil projects at 15–25% or more is seen as a means for extracting the developing world's natural resources on unequal terms. If only a small share of market capitalisation could be shifted to developing states, so the argument goes, most world poverty would be removed – though how such a transfer might be achieved remains to be explained. The unequal terms of trade discourse goes back to the 1960s and dependency theory. In reality, the corporate oil effort is diffused to the developing world via investments and reflected in joint ventures with a range of impacts outside the companies themselves.

Shareholder value in oil is treated simply as part of a zero-sum game, its price being the impoverishment of the exploited masses. Capital accumulation is associated only with high-concentration merger and acquisition activity and job-shedding – never with benign foreign investment. The "ring" of Marxist economic theory can be heard in these refrains. From this parody of economic reasoning, corporate oil emerges as an unrelentingly unjust and wicked force in the world. Why, if corporate oil were to fail to produce a competitive return, capital would not simply flow elsewhere – and why, if it did, that would be desirable – is not on this agenda. The old left's loss of intellectual traction following the Cold War is here translated into new arenas. Corporate oil has become one of the recipients of this flawed set of theoretical ripostes.

Similarly, globalisation is credited with benefiting only the top 20% of the world population, the richest mostly. Developing-country interests, meanwhile, are said to have been effectively silenced by indebtedness, dependence on the IMF and IBRD, and the need to gain the good will of the elites from the North. Selective facts of world poverty are often used to make a compelling case. Corporate oil's presence in any poor society is almost invariably treated as the originating cause of poverty. All those who benefit from oil are thereby – indeed, almost by definition – tainted. Only those who oppose it, notably the NGOs, can be pure.

It is no exaggeration to say that corporate oil is demonised in the mythology of anti-globalisation. The movement is rarely concerned to make the kind of significant distinctions between companies that are evident. All oil companies tend to be tarred with the same brush. Indeed, since global economic competition is characterised as a "war against the majority", no distinction is made between oil companies and dictatorial military regimes: all are seen simply as allied machines of exploitation. Moreover, oil companies are charged with co-operation with assorted elites, often depicted like the mafia. Indeed, the industry is presumed to be fuelling, both literally and metaphorically, all sources of evil – not only global capitalism, but also imperialism, fascism, colonialism and apartheid (the final term given a broad definition, to include not only South African history but also the predicament of so-called Palestinian bantustans, and even a "notion" described as "global apartheid").

The anti-oil agenda, broadly entwined with that of anti-globalisation, seeks to build an integrated vision with which to reshape world consciousness, rousing the world from what it takes to be a form of amnesia over the unjust conditions out of which corporate oil alone is said to profit. In the process it seeks to unite diverse groups of activists, NGOs, alternative energy lobbies, political parties, workers, students, greens, dissident intellectuals and the dispossessed.

The anti-oil programme for nirvana is multifaceted and wide-ranging: it aims to capture the symbols of moral hegemony, shame the shameless (most notably, oil executives), force governments to respond with regulations and inhibit the operations of the oil companies. Its proposed means include "fair trade" terms, increased taxation on corporate capital, the closure of tax havens, withering dictates to guarantee transparency, the cessation of IBRD funding for hydrocarbons, heavy environmental regulation and enforcement of "sustainable" energy systems, with the cancellation of all developing-country oil-related debt. Plausible substitutes for an oil-free world are never indicated.

The focus of anti-globalisation varies between its adherents and members (there is in fact no criterion for this). Some stress the desire for a fairer balance between North and South and the need to reform institutions such as the World Bank; others want a more radical decentralisation of the world economy; some wish to restore the authority of the nation state combined with a more civilised capitalism or traditional socialism; and many stress the need for de-oiling (or even, more radically, de-industrialisation), with the future accent to be put on solar and renewable energy.

There are two fundamental points about this global anti-oil agenda. First, it is difficult to see – and it is never explained – how all this could be implemented without some form of quasi-Stalinist world state, with the great and the good of anti-globalisation as unelected and permanent members of a moral politburo. It is pertinent to note that much of the anti-globalisation movement has its origins in earlier Marxist and/or leftist ideas that have encountered strategic defeat. Yet some see the shift to resource nationalism and American decline in oil-world prominence as evidence of triumph for the old ideas.

The second, and perhaps even more important, issue is that the anti-oil agenda is forever expandable. Oil companies that respond positively to demands from these lobbies attract further allegations and demands. There is no logical reason why this process should ever end, except perhaps with an "end of oil" syndrome. While companies such as Shell adopt the mantra of "sustainable development", so the proselytisers of that term are free to revise and expand the criteria that define such development. Ever-changing criteria in any case result inevitably from the diversity of a movement that is more united in what it is against than what it stands for. A game in which the goalposts can and will shift, and where there will be no final whistle, is not one that corporate oil can win.

In the West, the anti-oil movement and critics of oil companies have generally been non-violent. Their main methods have been symbolic and ideological. They have succeeded in influencing mainstream debate. Think of the way that, for example, George Soros has enthusiastically supported demands for transparency in oil deals. The movement is likely to recruit many more glitterati from politics, the arts, finance, Hollywood, academia and elsewhere in future. Campaigning in the West has had some low-level effect on corporate actions. In parts of the developing world, violent action against corporate oil is more frequent. Whenever it occurs, there seems little shortage of commentators to justify such action on the grounds that, though it is far from pretty, it is not as ugly as the "structured violence" of global capitalism.

Despite the fact that, according to its own demonology, it might face incalculable odds, much of the anti-globalisation movement is optimistic about its prospects. Many anti-globalists believe that they have the international institutions on the defensive. Certainly Davos is now responsive to the World Social Forum and its arguments. Institutions such as the IBRD, through its Global Compact, and the UN have sought to engage many NGOs in dialogue. The rise of this fifth estate on the world stage reflects their success. Even so, many anti-globalists argue that the preference shown to some NGOs over others is an attempt by these institutions to divide the movement and hence rule over it.

For populist movements, anti-globalisation has certain advantages. It offers a refuge and a platform for those older campaigners bereft of the now-defeated ideological movements of earlier generations. To the young it offers the appealing promise of a cure for the corruption and despoliation of the world. Its slogans – to the effect that everything is the fault of corporate oil – are easy to grasp, remember and disseminate. They even provide a viable business model.

The myriad contemporary struggles between corporate oil and anti-globalisation appear likely to intensify. Already the contest is played out in multiple arenas, including commercial, legal and organisational milieux, and in regional, country-specific locations, even project-specific playing fields. As the search for oil broadens, so the less desirable, more extreme locations – the failed and failing states, the countries run by despotic regimes, the illiberal worlds – become even less avoidable. This will fuel the intellectual and activist fires found in world oil debates.

The oil industry has learnt that it needs to respond to the anti-oil movement, or at least some of its ideas and claims. The dangers, after all, are clear for all to see. Statutory regulation and redistribution, for example, hardly encourage unconstrained investment in oil. A world of massive underinvestment in hydrocarbons would suffer dramatic drops in growth and prosperity. Whatever the merits or not of oil, alternative energy has yet to provide the solutions that the world needs in affordable energy, or sufficient profitable opportunities desired by the environmentalists.

Lord Browne, when chief executive of BP, led the move towards company engagement with NGOs, arguing that corporate oil needed to look at such entities as potential partners in dialogue, rather than as implacable enemies. This will be harder than imagined with many of them, and yet the ideas drawn from anti-oil bodies remain potent in the minds of many public audiences.

Curiously, however, corporate oil seems reluctant to engage with and debate many of the core issues (perhaps it is like discussing its own execution). Most importantly, it shies away from playing its trump card. That is, it refrains from arguing that, on portfolio cost/benefit terms, the oil industry and oil companies can usually be shown as huge net contributors to global living standards and world development.

Instead, the emphasis in much corporate response has settled for well-worn mantras and somewhat tired strategies that suggest to the public the undiluted merits of corporate social investment and sustainability – whatever that might mean. The industry has also lobbied for its interests inside the corridors of power with governments, and it has an imperfect understanding of its modern nemesis: the modern barbarians.

✳

One of corporate oil's most common defences deployed against barbarian threats, both traditional and modern, is "corporate social investment", conducted under different guises by the industry for several decades. The term is a broad one, encompassing education and training, community development, health care, welfare programmes, enterprise development, and cultural, artistic and recreational activities. The industry's hope is that such programmes will deflect, disarm or pre-empt criticism and even create a sense of common interest with selected local stakeholders. It may just empower corporate oil's opponents and adversaries.

Such strategies have been used most directly as an attempt to shore up the support, or at least the tolerance, of oil-proximate communities, especially in the developing world. Social investment has also sometimes included a biddable element with which governments influence and select oil companies during contract negotiations.[80] In effect, social investments act as a form of security and sometimes as a means to lobby for local industry requirements. It is noticeable that foreign players from outside

Western oil typically indulge in much less of this practice and are rarely mandated to do so by their own states. Whatever the ethical or moral claims made, this has often acted to corporate oil's general disadvantage.

Typically such programmes are supported with public relations activity. On occasions glossy advertisements and associated public relations advertisements consign hydrocarbons operations to the background or even omit them altogether. One of ENI's advertisements for Karachaganak oil and gas, for example, depicted a sleigh with huskies on a snow-swept wilderness of pristine beauty beneath blue skies. It told us: "This is the ENI Way". It showed nothing of the oilfield itself. Similarly, Shell used a litany of print and television advertisements to portray its forward-looking, green credentials: "We listen with particular care," said one, with an image of a fisherman in a trawler with a sunset behind, adding that, "Exploration … should benefit everyone and harm no one." The Ogoni people of Nigeria might not have bought this spin. Again the state oil competitors to corporate oil and many local companies in the developing world do not follow this benign advertising pattern, which raises the question: To whom is it directed? Almost always, it seems, it is for corporate oil's own preferred audience.

Other advertisements found in the industry often extol the virtues of the company's social investment programme, usually with photos and statements of good intent. The question here is whether these types of advertisements – there are many of the same genre – accomplish anything substantial. They may just deflect focus from a more realistic portrayal of industry interest and the full spectrum of benefits that oil may provide as a key component in both modern societies and the developing world. Corporate oil may be selling itself short in its public relations pitches. National oil companies have also engaged in some ambitious public relations and are not immune to the same myopia, but most keep a sharper focus on their primary core interests.

The phenomenon of social investment stretches over most of the length and breadth of the industry. So does the litany of justification. It is a sort of parallel universe in which the battle is fought over ideas, images, good intentions and implied virtue. Perhaps it assists in intercompany competitive positioning with Western interests: it has little or no impact on the mindset of the anti-oil lobbies.

The very notion of social investment programmes induces such a warm glow that we might wonder how anyone could question whether they are not a "good thing". No doubt many projects do indeed provide some benefit in their own terms. But an assessment of the role, value and advisability of social investment is far from straightforward.

Just as Machiavelli in his worldly analyses of social behaviour looked past expressions of piety and asked which interests were being served at the time, so we should put the feel-good factor to one side and examine whose interests are served by corporate oil's social investment programmes. The answer – to the surprise of their critics and, no doubt, to themselves – might not be corporate oil's. Indeed, there are many reasons for scepticism over the wisdom of corporate oil's recourse to social investment and dependence upon it as a means of salvation.

First, the concept of social investment is prone to fuzzy thinking and hence much confusion. What is a social investment project: an ineffective philanthropic exercise, mere public relations, or genuine investment? If it is actual investment, what kinds of return will it produce, how can it be measured and to whom will it accrue? It does not help that some of the answers rely on terminology that is itself fuzzy.

Fuzziness here produces two risks – that of conflicting expectations, as different parties perceive projects differently; and that of multiple, mutually incompatible project management objectives. In and around the oilfields of the developing world, the practice creates division between those advantaged and those left aside. Governments are also enabled partially to abandon their public responsibilities, and in some cases the process can be corrosive to the integrity of the states within which they are conducted, as fickle corporate largesse replaces public obligation.

Second, there are the usual problems associated with appeasement. Meeting local criticism of the social costs of the oil industry by making, in effect, compensatory social investment is likely to raise the stakes in two ways: it encourages the original critics to make further criticism in the hope of gaining further recompense; and it may encourage new critics to enter the arena in the hope of obtaining a piece of the action. Social investment as a form of appeasement is, therefore, almost always at least potentially self-defeating.

A further difficulty is that although social investment may solve some local problems, it may cause others. One set of problems concerns payments. Some projects are both funded and managed by oil companies, but others are funded through corporate donations to third parties. Where donations are involved, there are potentially problems over probity and transparency. This is especially true when the third parties are tied to governments, as can be the case in Angola and elsewhere.

Another type of problem often caused by social investment is that of dependency, as has been the case with aid. Over time, governments may be tempted to shirk their responsibilities and rely on the oil companies to pick up the pieces. When an oil company enters a clear contract with a national government and pays its royalties, tax and commitments under contract, it should then ideally be the responsibility of central government to provide the necessary social investment. Where this has not been done effectively, as in Nigeria and Indonesia (where there has been decentralisation, with different provincial tax authorities emerging), oil-proximate communities are often left to rely on corporate social benevolence – or to threaten facilities with sabotage or personnel with kidnapping and violence, unless and until such projects are introduced.

Yet another problem with social investment is that it involves oil companies in a game that is unwinnable. For all its immense resources, corporate oil simply is not able to "fix" all the world's problems. To alleviate all poverty, to end all conflict – these are tasks beyond the capacity of any single industry. Social needs will always exceed the capacities of corporate budget allocations to such assigned tasks. Each time one social problem is solved or alleviated another problem is waiting in line. This in turn necessitates choices to be made over which concerns to prioritise – a process that risks stoking resentment over unfulfilled demands or expectations. Thus corporate oil's success with social investment is in the long run almost always only temporary. The success of particular projects may provide attractive photo opportunities, producing attractive images for corporate websites and annual reports. It cannot, however, conceal that truth that corporate oil has entered into a process that has no visible end-game.

Closely related to corporate oil's social investment strategy are its sustainability programmes. "Sustainable development" has become a fashion, lauded like apple pie. It is one of those things most people in Europe and the US like the sound of, a common denominator for agreement. The phrase has the great advantage of seeming to combine a concern with the needs of both the environment and society. Discussing sustainable development at board level seems to provide an opportunity for corporate therapy. Even oil executives wax lyrical about company ratings on Dow Jones Sustainability Indexes and attempt to link them – not too convincingly – to share prices. The phrase is also helpfully vague: there is no consensus on its definition. It is, therefore, eminently usable in press releases and annual reports. Its malleability makes it a spin doctor's dream – and indeed a politician's, since the phrase can be used to support all manner of policy initiatives without providing any inconveniently clear criteria on outcome or accountability.

The concept involves very real difficulties. The requirements of sustainability are not always readily reconcilable with those of development. Where there are awkward trade-offs to be made between the two, "sustainable" can become a synonym for "slow" or "retarded". To some, this may even be part of its appeal. For developing countries there is much aversion to its practice. Moreover, a call for sustainability rapidly transforms itself into a call for greater regulation. New regulation not only carries a cost in itself but may also benefit companies from territories outside the regulatory framework, typically foreign state oil competitors.

The idea is problematic in other ways too. It is a relative concept, since what is "sustainable" for one party may not be for another. This is pertinent to the programmes of the oil industry. Some social programmes, after all, are unable to attract or generate funds to cover future recurrent expenditure or maintenance costs. They can in those cases "sustain" themselves only by the further injection of capital. Moreover, the upstream world is globally mobile: the industry must move on as some sites are made redundant and new ones develop, or even where companies exit a country as sometimes they must. There is, therefore, inevitably a risk of churn in so-called sustainable projects.

The concept of "sustainable development" raises the question of what constitutes sustainability. Assessments of sustainability invariably rest on wider assumptions about the world's demography and multiple environments – whether or not those assumptions are made explicit. One notable example of such an assessment is the report produced by the World Wildlife Fund (WWF) ahead of the Johannesburg Earth Summit in 2002. It ventured the opinion that, if current consumption trends continued, the world would "expire by 2050" – a sort of Malthusian solution. A definition of what is sustainable based on the WWF's assumptions would obviously be highly restrictive. Indeed, according to

Friends of the Earth: "Oil companies should be planning for the day they get out of the fossil fuel business rather than constantly trying to expand markets for unrenewable and climate-damaging petroleum." Such assumptions are, however, open to challenge. A much more empirical view of what constitutes sustainability emerges, for example, from the trend analysis to be found in Bjorn Lomborg's *Skeptical Environmentalist*.[81]

Oil companies such as Shell, BP, BHP Billiton and Neste Oil engaged in the World Summit on Sustainable Development (WSSD), held in Johannesburg, offering their own assessments and projections. Shell's Exploring Sustainable Development led the way with its scenario planning. There is of course no agreement on the most likely scenario. Differences arise from the disciplinary frameworks and methodologies used, the ideologies embedded in the forecast, the inconclusiveness of scientific data, the views taken on a range of specific issues (including Peak Oil, the potential role of alternative energy sources, the functioning of price and market mechanisms), and the scope for human creativity and elasticity. With such critical pliable variables involved, almost any outcome could be modelled and justified.

✻

Let us now stand back from the particular issues and consider corporate oil's strategy in the round. Overall, the industry's reliance on warm, fuzzy, social investment and sustainability programmes as a defence against barbarian threats possesses one huge flaw. It is that the strategy plays to corporate oil's weaknesses rather than its strengths. According to Sun Tsu in the ancient classic *The Art of War*, this is an error: "Those who are first on the battlefield and await the opponents are at ease; those who are last on the battlefield and head into battle get worn out."

Oil companies are, after all, neither governments nor welfare services nor collectively some sort of world development agency. They are in the business of producing and selling hydrocarbons. That is what they know and do best. The public the world over senses this instinctively and, for the most part, recognises that the world needs efficiently developed and produced oil, gas and energy. Corporate oil may be ill-advised to neglect its central roles and instead, as is too often the case, spin softer, benign images of its world that can so easily be contested by its antagonists.

The growth of concern with social investment, sustainable development and political correctness has produced a distorted business climate for corporate oil. This in turn produces some culture change inside the companies. They find that attention can no longer be given so exclusively to the bottom line. Above all, there is a new emphasis on caution. This stems from a number of sources – not only self-selected public relations sensitivities, but also greater surveillance from NGOs and demands made subsequently for more regulation.

Additional burdens on companies have arisen from the demand for transparency over payments made to governments. The call has come from dozens of NGOs (notably Global Witness, Amnesty International, Christian Aid, Friends of the Earth and Oxfam) and has been endorsed by George Soros. The initiative perhaps shows a rather touching faith in the efficacy of rules. These are unlikely to stop such long-established traditions as "wasta" (the use of friendships and reciprocal favours) in the Middle East or "dash" (small payments to lubricate transactions) in Nigeria. What has been regarded as corruption in one culture may be seen as a commercial norm in another. Moreover, rules often succeed more in changing the forms or loci of corruption than in removing it altogether. It is not surprising, then, that some of the world's most corrupt governments have had no qualms over introducing anti-corruption laws. There appears, in any case, no adequate published evidence that questionable deals or payments in oil have ever amounted to more than a small fraction of the value of all oil deals or payments worldwide.

Corporate oil faces a new, politically and socially "correct" business climate. This in turn is producing a more conservative, rule-based, edict-bound, puritan culture within some companies. It is a culture in which new initiatives are all too easily dismissed as "against our principles", "not in harmony with our code of conduct", "not the way we do business". The new internal and external climate of correct behaviour influences the whole dynamic of many oil companies. At the top, there may be the temptation to promote the staid and the stolid at the expense of the deal-driven executives in the oil circuit. Despite efforts to increase the diversity of senior management, the outcome may be greater homogeneity of thinking. There is a temptation too to select non-executive directors in oil companies on the grounds of acceptability rather than capacity to add value.

The cumulative effect of these changes may be to shift corporate oil closer to the phenomenon of the "nanny-style" oil company, akin to the phenomenon of the micro-managed nanny state in which a cradle-to-grave risk-free existence is promoted and sought. Corporate oil might come to look a little more like the soft, benign images with which it promotes itself in advertisements and communications. All this might be welcomed by onlookers from the NGO world and even from certain shades of government. But, as Machiavelli would no doubt recognise, things are not so simple and the barbarian worlds seem less likely to take this course. However much the complexion of corporate oil may change, the old challenges remain. The new ones will be even more daunting.

✳

The management of portfolio in barbarian worlds is likely to remain an uphill struggle. Corporate oil has typically made liberal use of lobbies in its own domains to smooth the path for its interests. It is less accustomed to or informed about the modi operandi and conventions to be able to do so in barbarian worlds. Lobbying itself also draws criticisms from its opponents. An understanding of these phenomena is instructive in appreciating corporate strategy.

Corporate oil is much more used to dealing with acculturated political elites in very different environs from those of the developing world. This is, after all, the way the West itself operates. In America, for example, corporate oil is an important source of finance for politicians and their selected parties. The large companies make political donations (openly and legitimately) of tens of millions of dollars (within the US over three-quarters go to the Republicans). Many senior figures in the US government –

Dick Cheney, for instance – have moved between oil and politics during their career.

Close links between oil and political elites are not peculiar to America. In the UK during Tony Blair's tenure as prime minister some critics would jibe that "BP" stood for "Blair Petroleum". The signing ceremony for BP's $400 million deal with Sidanko was held at the prime minister's residence, 10 Downing Street. Although BP has not made donations to the Labour Party, there have been links at the level of key personnel. Some BP executives, such as Bryan Sanderson (former head of BP Chemicals), David Watson (BP group treasurer) and Alan Jones (then BPAmoco Scotland), were appointed to governmental task-forces. Sir David Simon, once a BP senior executive, was ennobled as Lord Simon and became Labour's minister for European trade and competitiveness. Anji Hunter, a close confidante of the prime minister when she worked in Downing Street, became head of internal communications at BP. Philip Gould, Blair's opinion polling guru, was hired by BP for its rebranding exercise. BP has also run country-house seminars for the British political glitterati. The oil link to politics is not unusual, exists worldwide and is found in barbarian worlds as well.

That close connections between political elites and corporate oil are widespread and well established is unsurprising. Curiously, though, some NGOs seem to regard such arrangements as peculiarly Western and thereby, it seems, both contaminated and contaminating. It is as if when Western companies seek to replicate such connections with governments in the developing world, the political systems there are somehow being poisoned by wholly alien practices. It is not always so.

Close relationships between oil and political elites can be found across all six continents. This is especially true in countries that have national oil companies. In South Africa, for example, appointments to CEF senior management and the board of Soekor-Mossgas (now PetroSA) are all made by the government and have often consisted of ANC nominees. In Malaysia, the CEO of Petronas has always reported not to the minister of energy, but directly to the prime minister's office. And so on – examples from across the world are not difficult to find.

Often the relationship between oil and government centres on the head of state. The financial outcomes are perhaps predictable. Many a leader in oil-rich Africa has not acquired personal wealth on a presidential salary allied to a uniquely high savings rate. Some in Central Asia have reportedly held accounts in Switzerland that were furnished with oil money from oil deals in the 1990s (in excess of $1 billion, it has been claimed in one case). Cameroon for many years ran a secretive *caisse noire* overseas that diverted oil funds offshore under presidential discretion. Successive Nigerian presidents have been enriched by oil funds. Ex-President Sani Abacha is reputed to have amassed funds overseas of $3 billion: these becoming the focus of repatriation by former President Olusegun Obasanjo. The House of Saud is wealthy and its many princes depend on a rentier state for their riches and oil-based largesse. In Brunei, the sultan's enormous assets have been built on oil and gas revenues operated by a state-owned JV with Shell. And (again) so on – these examples are far from exhaustive. By any comparison the ties between the Western political elites and corporate oil at best result in marginal benefits for most oil-connected politicians.

One of the most common forms of connection between the oil industry and government is the lobby system. Lobbying is much monitored by NGOs, who regard the system with distrust – perhaps forgetting that they are themselves lobbying groups.

In almost all major capital cities, lobbying – from all large industries, not just oil – has become a sort of industry in its own right. Sometimes companies lobby directly. Lobbying in private, however, can be counter-productive, enabling opponents to play on suspicions that the democratic process is being subverted. At other times companies use intermediaries, including purpose-built organisations. Most companies contract professional specialists in government affairs or international relations for the purpose. Corporate oil also advances its interests through bodies such as the Business Roundtable, International Emissions Trading Association, International Petroleum Industry Environmental Conservation Association (IPIECA), International Chamber of Commerce (ICC) and World Business Council for Sustainable Development. Other third parties that act as intermediaries include think tanks, research institutes and academic experts.

It is instructive to consider, as an example, the formation of policy in America concerning African oil. When in 2002 the African Oil Policy Initiative Group (AOPIG) produced a report on African oil reserves and American energy security, it recommended that the Gulf of Guinea be declared a "zone of vital interest". It proposed that America should establish a specific military command for this region and a naval base in São Tomé & Príncipe. The working group included some members of the Bush administration. A formal lobby association (US-Africa Energy

Project) with extensive links to the oil industry was established. Contributors to the promotion of this venture included Anadarko, BP, the then ChevronTexaco, Kerr-McGee, Marathon, COP and Shell – all companies that were then active in the Gulf of Guinea area. None of this is illegal or untoward.

The National Intelligence Council, associated with the CIA, has unsurprisingly taken an increasing interest in African oil. It expects some 25% of US oil imports to come from Africa by 2015 (compared with around 15% now – more than Saudi Arabia at present). The Corporate Council on Africa has likewise been active in promoting oil industry interests. Canada is setting up an equivalent, the Canadian Council on Africa. In the UK, UK Trade & Investment has groups that assist companies that focus on Africa, the Middle East and elsewhere. So do most Western states. In some cases such lobbies are mediated by specialist firms.

Lobbying firms in Washington focusing on Africa (not only in relation to oil, but also to trade and politics more generally) are numerous. They include Ryberg & Smith, Valis & Associates, Patton Boggs and Cohen & Woods International. Transafrica, run by Randall Robinson, cultivates the Afro-American lobby's interests, as does ex-ambassador Andrew Young's Goodworks International (with a focus on Angola and Nigeria). Among relevant think-tanks can be found the Georgetown School of Foreign Service, Council on Foreign Relations and Brookings Institution. There is also the Center for Strategic and International Studies, a well-funded bipartisan body using top-quality scholars, with a focus on energy security and political risk and a global approach to world affairs. A host of new lobbying bodies has sprung up as the profile of African oil in American foreign policy has risen.

There is nothing peculiarly American about such activity. The issue of African oil has generated interest of a similar kind in the other major capitals of the world. Institutes in Moscow have monitored the development of African oil closely. A Russian "return to Africa" is already in motion, led by Gazprom and Lukoil. President Putin himself has already made forays into the African oil patch with visits to Angola, Nigeria, South Africa and Gabon. There have been similar efforts in China, culminating in 2006 in a major Sino-African summit meeting with most heads of state in attendance. Oil was high on this agenda and in private meetings held with African heads of state.

There are, then, a multitude of relationships, both direct and indirect, between oil and political power. They include, *inter alia*, links between corporate oil and Western governments, corporate oil and governments in the developing world, and state oil companies and governments. That corporate oil is well versed in such relationships, having long practised them in its own countries, does not guarantee that it will be successful in gaining preference abroad. Corporate executives are by no means guaranteed to be a match in future for the local elites of the barbarian worlds.

Likewise it is found that the developing-world oil milieu operates with similar lobby structures and political lubricants, even if these do not exhibit the institutional forms and conventions of Western finesse. In effect they are adaptations to local circumstances. Presidential offices are often the fulcrum for these endeavours, with ministers of energy a close second as key ports of call. National companies are in this matrix of influence, and extra-political links may be found in the social structures of authority. The locals understand and know these informal bodies and relationships well. They are often more difficult for outsiders and corporate oil to penetrate, manipulate and manoeuvre within. It should not be surprising that the locals, especially those with connections or who are members of an ethnic group or specific political party, have some edge in this part of the game.

＊

To see the real significance of shifting industry culture, we need to return to the stereotypical yet instructive images of barbarian and corporate executives earlier drawn. The changes in corporate culture and strategy identified might be the last thing that oil companies really need to confront their future. It could render Western corporate oil less well equipped to do business with its traditional barbarian opponents and the emerging empires of oil.

In many arenas the industry may become more vulnerable to state oil competitors from emergent empires of oil. They will be typically less encumbered with concerns over social investment etiquette, mantras of sustainability, demands for transparency, angst over lobbying, a forest of regulatory impositions and worries over what NGOs might do next.

In this last sphere, the complicated and complicating world of NGOs, corporate oil seems not to be very aware of the mounting pressures that could arise from this fifth estate and its growing involvement in the world oil patch.

I I

Corporate oil's modern enemies

Worldwide it has been estimated that there might be over 1 million NGOs, and whatever the actual count the number is growing. An internet search for "NGOs & oil" in Google produces over 1.3 million items. The NGO world represents a veritable fifth estate, one closely linked to the fourth estate of the media and press but distinctive in character. Though not all have engaged with the oil industry, many have, and more are likely to do so.

Although in cases derived from radical movements in previous eras, the NGO phenomenon is in many ways a post-modern one, even where some have a long vintage in this "market". The tools used vary, but typically include research, global communication and dissemination via the internet, social networking, inter-NGO coalitions, shareholder activism, legal sanction, moral pressure and a focus on so-called "soft" (but in fact not soft at all) targeting, such as the negative branding of oil companies, their products and reputations.

Through such means many NGOs, both individually and as a group, have had an impact on the oil industry. Often this has been purely in terms of publicity. In some cases, however, the hits have been more palpable. One well-known example is Global Witness's "naming and shaming" of companies involved in Angola over alleged corruption, lack of transparency, the misuse of signature bonus payments by the state, arms-for-oil deals and the financing of the now-settled civil war. Though no company exited Angola for these reasons alone, such action raised risk perceptions. Another hotspot for NGO activity has been the Caspian region, where many NGOs have been involved in issues concerning oil and the environment.

In this climate, many companies have sought to take pre-emptive action by making investment offsets in community projects or by engaging in public relations. The latter, however, carries its own risks relating to legal action concerning complicity and even comment. Public statements have been challenged in Californian courts, where they may be categorised as "commercial speech" and hence have a lower degree of constitutional protection than ordinary public discourse. Companies may find themselves being sued for "false and misleading statements". Incorrect facts or misleading impressions may lead to punitive damages. Corporate oil is vulnerable: for example, companies that gloss over inconvenient facts when justifying their involvement with impure regimes, or that present themselves as good corporate citizens without providing a scrupulously fair account of their activities, may find themselves facing prosecution.

Except where proxy wars were fought (as, for example, in Angola), the antagonists of the Cold War left each other's oil industries untouched. With the geopolitical fragmentation that has followed the end of the Cold War the industry faces a new array of sophisticated opponents. They include developing-country activists, environmentalists, student groups, human rights groups, gender groups, anti-capitalists, radical Christians, anarchists, socialists, peaceniks, unionised workers and a bewildering range of single-issue oil-related lobbies.

These groups have the advantage over corporate oil of being footloose, which, ironically enough, is the product of the very process of globalisation that many anti-oil groups oppose. In contrast to enemies of Western interests in the past (the Soviet Union, for example), anti-globalisation lacks a geopolitical base or substantial material assets. Corporate oil, in contrast, unavoidably has both. It therefore constitutes something of a visible and vulnerable target – just as in the Roman Empire the citadel of Rome once did for the barbarians of the ancient world.

There is no single ideological tenet shared by all modern barbarians. They draw on a wide mix of ideological traditions, notably fragments of Marxism, anarchism, dependency theory, syndicalism, autonomism, feminism, green thought and theories associated with climate change, alternative energy and oil-driven catastrophism.[82] Just as the ideological sources vary, so do their goals. They include the "de-oiling" of society, the de-industrial-isation of society, the greening of the world and the de-linking of oil from societies in conflict. The consequences of any of these programmes for nirvana are almost never explained.

It would be a mistake for corporate oil to take a monolithic view of these entities. Some NGOs are suspicious of, or opposed to, each other – especially after the G8 summit in Genoa, where violence from direct action groups was not supported by others.

Overall, the diversity of the antagonists presents a difficulty for corporate oil. The opposition is dynamic and its philosophies (containing elements of both myth and analysis) are elastic. Their diversity widens the spectrum of concerns, confuses debate, fosters claims to moral authority, and maximises the likelihood of ideological and in some cases even physical conflict.

Several oil companies have engaged in dialogue with NGOs in a number of ways. Some have even sought advice and consultancy from specialist NGOs over the management of social investment projects, though purists within the NGO movement frown on such co-option. Dialogue has often been productive. ExxonMobil's discussions with NGOs over its Chad–Cameroon pipeline did the company no harm. Dialogue does, however, carry its own risks. It can distract companies from their core business. Perhaps more importantly, it could result in a risk of litigation arising from the legal status of "commercial speech".

Oil companies rarely attempt to act in concert and must abide by antitrust laws. Such formal associations as exist within the industry are bureaucratic, inward-looking and even moribund – hardly the kinds of organisations suited to countering smart, flexible modern barbarians. Each company therefore constructs its own defence. "Defence" is indeed the operative word, since corporate oil has made little attempt at pre-emptive action. More pre-emptive strategies may be used in future, together with in-house counter-intelligence measures. If they are not, the grounds (literally or metaphorically) on which the companies can operate could be ceded to the morality imposed by the ethos espoused by modern barbarians.

The dynamism inside NGO models should also be recognised. They can elect, deselect, modify and transform the area of debate, and even the issues of concern. They therefore hold some potential to be a volatile force within an industry seeking stability. Yet a few commentators argue that the industry has become more stable over time.[83] Their focus, however, is more on market issues than on broader definitions of stability in and around the oil world patch, which in this author's experience has progressively become less stable.

In this book we have used the notion "modern barbarian" as an umbrella term to describe the antagonists or putative enemies of corporate oil and the older empires of oil found largely inside Western worlds. There are now, just as there were in the ancient world, many different types of barbarian. We should now chart this modern barbarian world, distinguishing its "tribes" and concentrating in particular on the NGOs, many of which have a benign agenda and some of which perform sterling work. It is useful to focus on both the geography of selected NGOs and their generic anthropology.

✳

Numerous NGOs now take an active interest in world oil affairs. They have between them all manner of concerns – including, inter alia, corrupt oil regimes, signature bonuses, oil-for-arms transactions, oil-for-food, oil depletion, selective divestment, rainforest protection, indigenous people's rights, Aboriginal welfare, depleted fisheries, endangered species, forced labour, child labour, oil and war, slavery in oil societies, gender and women's equity, transparency and governance, empowerment, and corporate social responsibility. The breadth of focus varies hugely between NGOs. Some are specific to localities, nations, regions, issues or targets – others are more general and international.

Some NGOs are interested in the oil industry only as part of broader campaigns. Others focus specifically on the hydrocarbons industry. Oilwatch, for example, which was founded in Quito in 1996 and now has member groups in over 50 countries, describes itself as "a resistance network that opposes the activities of oil companies in tropical countries". It acts as an information exchange on corporate oil's alleged transgressions on a number of issues, including sustainability, communal rights, biodiversity and human rights.[84]

Evidently there is no shortage of funding for many NGOs – whether from private donations, state support, or institutional sources. As their numbers grow, so too do the possibilities for networking and coalition between NGOs. The influence of the fifth estate as a whole has certainly been extending around the oil world. The IBRD and UN have entered into formal relationships with many NGOs, with around 1,500 logged as UN observers at the last count.

The actions of NGOs have on occasion had a direct impact on the financial markets, as well as an indirect influence through their effects on oil investor confidence and corporate portfolio choice. They also require increasing amounts of time and energy from the managers of corporate oil – a cost that is no less significant for being largely hidden. Globalisation loosens the grip on state power held by many national governments, so the influence of NGOs is likely to increase still further. This is a world that the oil industry will need to understand better.

Organisations involved in environmental issues form a particularly active part of the NGO world and have recorded some success in precipitating changes in corporate oil behaviour. Though some companies have reacted coolly – and questioned the scientific findings on which such campaigns rest – others have treated green issues (notably CO_2 emissions and climate change) as "hot button" issues. Already emissions trading mechanisms have provided new business for NGOs and profits for carbon traders, perhaps at a cost to consumers. Negative impacts on the environment, especially in Ogoniland and the Peruvian rainforest, have involved major players such as Shell in implied reputation risk, remedial and reparation costs, and more costly image management.

Green campaigns against oil have become increasingly international. Shell's venture in Sakhalin, for example, became the target for an NGO campaign including Pacific Environmentalists and Sakhalin Ecological Watch. They claimed to be acting on behalf of local interests, including fishing groups. Gazprom bought this particular project but – perhaps wisely – has retained Shell as the operator. The trend towards the internationalisation of campaigns is likely to continue.

Generally, the large companies have been anxious to establish their environmental credentials. They advertise their policies on diminishing carbon footprints and sequestration, invest in renewable energy and provide environmental audits in their annual reports. There has been a growth in more or less formal regulation, including self-regulation, in relation to such issues. Companies certainly do not enjoy the freedom to decide on their activities unfettered, as is well illustrated by Shell's withdrawal from Camisea in Peru on economic grounds, but after negotiations with over 100 NGOs on issues such as the rainforests, indigenous rights and environmental impact. The last lobby, purporting to act on behalf of nature and humanity, claims for itself a position of moral virtue. Many NGOs are not slow to exercise a well-known art of management: taking the credit themselves and allocating the blame to others. Corporate oil appears to be an ideal victim.

In America, NGOs regularly call for sanctions against problematic regimes, as do some Western governments. The application of sanctions is a risk for all oil players, but especially for American companies. Sanctions not only limit the area they have to operate in, but also allow players from elsewhere that do not abide by the sanctions to gain competitive advantage. Though the US has long been considered a blue-chip environment for corporate oil, exploration and development capital has steadily been deserting the country in recent decades and targeting assets abroad instead, except perhaps for the Gulf of Mexico and Alaska. For corporate oil, foreign investments make sense: the US is a relatively mature play, contains closed areas such as the Alaska National Wildlife Refuge (ANWR), has been fertile soil for the growth of oil-hostile lobbies, and has developed an increasingly politically correct, regulated, culture that affects corporate oil's operating interests. Although the demands of energy security may provide some respite, NGOs in the US and elsewhere are increasingly training their firepower on corporate oil targets abroad.

The case of Alaska, in particular, is dramatic. The excluded areas of the ANWR and a southern region of Alaska contain what, according to official estimates, are significant reserves of hydrocarbons. Successive federal governments have locked away acreage equivalent in size to the eight states around the Great Lakes. Around 450 million acres have been subject to "stringent land-use regulations", making them all but excluded from exploration. A 1997 federal government assessment estimated conservatively that off-limit areas and heavily restricted lands hold between them some 47 BBLS and 381 TCF potential. Despite the restrictions, Alaska now supplies 17% of US domestic oil. The Trans Alaska Pipeline System (TAPS), which is critical to supply, has over its 25-year history suffered 50 shootings at its structures. If one were to take place in winter, the event could shut down the oil supply for a significant period. Danger thus lurks in and beyond the snowfields.

Green pressures and voter issues in Alaska have created a great divide between corporate oil and the domestic anti-oil lobbies. This has constrained activity to current ventures and thwarted significant new development in America's leading domestic oil zone (which has the potential to replace some crude import volumes). High recovery volumes depend on sufficient break-even prices and positive impacts from applied technologies and future exploration – but such intra-industry issues often prove too arcane to triumph over ideology in the heated debates over oil. Many reduce such debates to a simple binary opposition of polar oil versus polar bears.

Issues concerning indigenous rights have also been used to

block corporate access to and exploration of reserves in North America. Inupiat Eskimos filed lawsuits to block offshore exploration in the Beaufort Sea and in 1999 Alaskan natives lobbied BP to stop its $500 million Northstar project. Some locals and natives around Prudhoe Bay, however, are adamant that oil has brought new opportunities for adjacent communities. Divisions in the local population over oil issues often define social liaisons and affiliations.

The size and number of opponents of corporate oil ventures in America are impressive. In relation just to Alaska, their ranks include Earth First, Greenpeace, Sierra Club and the WWF, among many others. Significant lobbies also exist in non-Alaskan areas within the US. However, the largest growth in NGO concern inside the US has been over oil operations abroad.

The range of countries with which American NGOs have been concerned has widened considerably. Churches and Christian groups (including radical fundamentalists) have been particularly active. Issues concerning Arab oil, anti-Saudi interests (especially after 11 September 2001), and regimes such as China, Cuba, Libya, Sudan, Iran and Myanmar have proved particularly challenging. There are also country-specific oil sanctions lobbies (for example, on Iran and Cuba), Black Caucus interests regarding Africa and slavery, and so on. As American corporate oil seeks to exploit hydrocarbon reserves from a wider variety of locations, this NGO activity is likely to grow. The unpredictability of such opposition and the likelihood that it will become more virulent constitute risks that corporate oil must face.

European oil has not been immune from political disorder,

especially since the fall of the Berlin Wall. In particular, there have been aggravated conflicts in Bosnia, Kosovo, Montenegro, Macedonia, Albania and Serbia. The enlargement of NATO and the EU has made for new divisions in eastern Europe and uncertainty in non-EU territories such as Belarus, Ukraine and Moldova. Europe and Russia are forced to cohabit in a post-modern space in which questions remain about Russia's capacity to adopt the type of "rules-based" order that has become familiar and institutionalised in the West. The growth of dependence on Russian hydrocarbons makes Europe more reliant on Russian concord and stability.

Corporate oil has long had to deal with hostile critics in Europe, most of which emerged post-1973. They include environmental activists, diverse NGOs, and politicians representing coal and renewable lobbies. The industry now faces increased regulation from the EU and has had to deal with pre-emptive changes in tax and regulation in the UK North Sea, as well as requirements for renewables in the energy mix. The UK is keen to ensure that current annual rates of oil production and investment are sustained for as long as is feasible. The new "use it-or-lose-it" strategy on licensing will probably benefit some players, but many companies are likely to quit this region. Some larger companies have already staged an exit. This forms part of the global trend of oil capital away from mature zones with high costs and taxes towards frontier regions in the developing world, sometimes controlled by more daunting state powers.

There are now thousands of NGOs in Europe, many of them well-established and well-funded, and the number is growing.

Many cases of NGO ultra-activism exist. Greenpeace's Atlantic Frontier Campaign with its "No to New Oil" slogan is an example. This campaign, based on concerns over marine life and climate change, is opposed to oil production, floating production storage and offloading (FPSO) or seismic activity. The effect will be to raise corporate oil's costs and make exploration a riskier game from West Shetland to the Rockall Trough/Plateau and the Irish offshore.

European NGOs have targeted, in particular, loans to energy ventures from the EBRD. They dislike its support for nuclear energy, object to fast-tracking of oil ventures, criticise perceived lack of environmental procedures, oppose funds for Komi Arctic oil because of pollution risk, object to oil contingency plans (as at Sakhalin-2) and want to see more public consultations with greater emphasis on energy efficiency.

NGO demands for compliance often expand endlessly. The requirement placed on companies to debate projects with "civil society" and nominal mandarins (including self-appointed ones) adds to management costs. Many of the NGOs based in Europe are concerned with the oil game in other parts of the world as well. Some have policies directed against corporate oil's material assets as well as its portfolio choices.

Europe, as a result, presents corporate oil with distinctive strategic choices between operating in mature areas under tough taxation and regulatory regimes, managing the challenge of increased NGO involvement, shifting towards newly incorporated EU members (some still hostile to foreign oil players), or exiting this zone in favour of new ventures in the alluring but problematic developing world. Many have taken the latter path, to find new pastures but in still unreconstructed states. Even so, Europe retains a vibrant upstream industry, especially in those regimes with moderate operating environments.

Any geography of NGOs should not just indicate their influence on those found in the anchors of the modern Western world: the US and Europe. Such entities exist in all OECD countries, and an increasing number now sprout from the soil of the developing world. Meanwhile, links between those existing inside the realms of modernity and those located in pre-modern societies have been augmented.

✳

There is no definitive anthropology available on NGOs in the oil world. But some inkling of their origins, strategies, targets, links and modi operandi can be gleaned from a look at selective entities.

It is difficult to draw firm lines between different types of NGO, especially since in practice they sometimes work together. In general, however, there are three main kinds of NGO concerned with oil: the anti-globalists opposed to corporate oil; those opposed to American imperialism, including the role of corporate oil; and those that seek to reshape, redirect or reform oil company strategies, portfolios or procedures.

The NGO industry – yes, it is one, a sort of informal or quasi-quango parallel to unelected government – has become increasingly well funded. It often makes effective use of global fundraising campaigns via the internet. It also receives donations from wealthy benefactors and companies, even governments – in spite of the costs to them of some NGO "success".

The Network for Social Change has established a sizeable funding network among individuals each with a net worth of at least $350,000 and with $3,000 to commit per year. Key donors include Red Hat, the JMG Foundation (based on Sir James Goldsmith's bequest), the CS Mott Foundation, the Rockefeller Brothers Foundation and the Samuel Rubin Foundation. Some business groups such as Frontiers of Freedom (funded in part by oil companies) have, in response, sought to restrict the funding of anti-globalisation groups. For example, there has been a campaign to remove the tax-free status of the Rainforest Action Network.

Here we should note that, whatever the genuine concerns that lie behind the founding of NGOs, there is always – as would not have escaped Machiavelli's gaze – the likelihood that they will become self-perpetuating businesses in their own right, each with the need to score success in the oil patch in order to secure further funding and perpetuate its continuity. In this context, corporate oil is an attractive target. It is global; it is a "dirty" business; it is eschewed by the glitterati; and it is forced to do business with many unsavoury regimes and leaders. It is easy to link corporate oil to environmental and human rights issues. The oil industry is, therefore, likely to continue to feature as a target in campaigns designed to appeal to liberal society.

Moreover, NGOs are practised at securing support from respectable quarters – church interests, ethical investors, institutions and so on. Modern NGO man and woman are more than likely to wear corporate suits and come armed, not with a Kalashnikov but with a well-rehearsed ideology, which may at times represent its own form of risk to corporate oil – a threat that the industry as a whole is not always sure of how to confront.

The oil industry with its traditional and oft-required confidentiality-cum-secrecy concerning contracts, deals and arrangements has naturally attracted increased critical attention from well-organised entities such as Transparency International, Christian Care and Global Witness. This debate is enlarging as time goes by and as pressures grow on companies to conform to NGO demands. Several companies have now joined the EITI process, possibly as a means to secure political cover. Many others stay away from such issues and the forums in which they are discussed. The same schizophrenia is found among countries that produce oil: many remain outside the process, and perhaps some join as a political act.

The industry has been forced to improve its governance record over the years, though this has not yet occurred in all parts of the world. Only a few companies have earned deserved reputations for flawed oversight or malpractice. These cases have tarnished the global industry's image. Almost all companies have been punished in the court of public opinion for the transgressions of a few – a point not understood by the wider public. Contrary to popular perception, political payments as lubrication for illegal deals have not been the standard practice of corporate oil deal-making. This practice is "outlawed" by, for example, Shell (but was evident in the Enron case). Political contributions are not illegal in the US, where they have been extensive and have covered both sides of the street in a form of "one-way bets" to ensure access (and maybe even future assistance). In the developing world, pressures are seen to be growing, especially on local independents, to assist political parties and vested interests. Most companies typically and carefully steer away from such indulgences.

Yet bribery and corruption are some of the elements associated, rightly or wrongly, with global oil companies, and in the developing world the refrain will be heard that bribery is a two-way street. Shell now declares a no-bribery policy, even for small facilitations. It has practices for identifying breaches and stamping out intermediary "assistance" in deal closure. Only a few cases have been identified and all culpable employees dismissed. This policy applies also to Shell's contractors. Even so, it is most unlikely that all corruption will simply disappear from barbarian oil worlds. Corporate oil will receive much blame, rightly or wrongly, for this deficiency.

It is not a crime to lobby for corporate interests, but a backlash against this practice is emerging, especially where it has yielded untoward results. Super-majors and independents are active (legitimately so) in lobbying by way of intergovernmental bodies and industry associations, and even with governments directly. Some support the UN Global Compact, Global Sullivan Principles and Transparency International rules. The emphasis commonly found within the industry is on truth, transparency, fair competition, open dialogue, visible corporate interests and full disclosure. Most oil companies are acutely aware of such strictures, if only to protect their positions with regulators, stock exchanges and the equity markets.

Nonetheless, the push for principles has not eliminated some bad practices of the past – backroom deals, illicit payments, the search for insider edge in contracts and the like. The link from corporate oil to the realm of Caesar is an umbilical cord that is unlikely ever to be entirely broken. Critics will consequently always have an assured supply of issues on which to attack corporate oil.

The truth is that corporate oil has a huge impact on economic development through its portfolio impact on revenues, state budgets and foreign exchange. In Africa, the oil and gas industry has generally done well for the continent, and is set to do better, even though in many environs it remains largely disconnected from the human condition of the continent's populace. This is not primarily the "fault" of corporate players. It is the government's responsibility to manage the industry and ensure wise use of its financial gains to generate beneficial impacts. Even so, more lobby groups and NGOs in future will probably contest this issue. Poverty will increasingly be ascribed to the actions of corporate oil even where its real causes are historic (predating oil), structural or embedded (in country economic configurations), and often directly related to flawed state policies. Indeed, as much acute poverty is found worldwide in countries that lack any upstream oil industry. Many inside the NGO industry prefer, however, to ascribe blame for almost all the sins of underdevelopment to one culprit: corporate oil. The target is easily identified and can be readily linked to poverty by mere association.

Similarly, there is a strong focus among NGOs on all manner of stressed environments, now a matter of worldwide concern. These are both global and local in nature. They attract a huge number of NGO interests, and embrace a wide range of issues concerning shipping, oil spills, effluents, waste management, water impacts, fishing, the rainforests, conservation, endangered species, threatened eco-systems and biodiversity. More single-issue and generic groups are likely to confront corporate oil over such issues. The oil industry will in future probably stand accused of endangering an even wider range of flora and fauna, notwithstanding the facts and complexities of science or local conditions.

Yet of all the issues raised, it is in the arena of corruption that corporate oil is tagged most with the sins of commission and omission, in the eyes of many NGO interests and their wider publics. Concerns over corruption and the need for financial transparency form a strong focus for NGO involvement in and around the oil industry. Investors representing billions of dollars held in funds invested in oil equities have been persuaded to call for action on transparency across the industry.

The number of NGOs active in this field has increased and they sometimes work effectively in coalition. They have had some successes in engaging CEOs and international institutions in discussions on disclosure, oil audits, bonus payments and in-country financial accounting – sometimes under the auspices of the EITI. Corporate oil has been reluctant, however, to debate these issues in public, let alone to take the lead in pushing for mandatory regulation, a strategy favoured by NGOs.

There has been no shortage of agenda items. Government practices in Equatorial Guinea, for example, have occasioned a great deal of debate. There has been critical attention from the

IMF over the issues of transparency and the control of oil revenues in general. The whereabouts in particular of state funds once held in a Washington account at the Riggs Bank at the time caused a furore. CBS and, in the UK, Channel 4 both broadcast documentary pieces accusing the regime of not properly managing public funds. The government now says in public that it supports the transparency process. There have been numerous other allegations of transgressions elsewhere – in Gabon, Kazakhstan, Angola and especially Nigeria.

In Iraq the post-invasion Coalition Provisional Authority (CPA) sought to pre-empt such debate by agreeing to publish full accounts on oil funds and spending. The Nigerian government says it will disclose all oil revenue receipts from oil companies. This commitment is echoed by Angola, where the government is keen to unlock aid monies available for use in the country. São Tomé & Príncipe published all details of awards and bonus payments received by the Nigeria–São Tomé Joint Development Authority (NSTJDA). Congo-Brazzaville was required to accept an audit and provide certification of its oil receipts in order to qualify for IMF support. The IMF is also seeking to terminate the use of oil-backed loans, favoured by several banks, based on the mortgaging of future production and the use of tax havens for the dispersal of funds after trades have been executed.

Such issues have been at the centre of the debate on the EITI process initiated by the British government. The oil companies (including ExxonMobil, COP and Chevron) want the emphasis to be placed on the governments and any code or compact to be voluntary. This suggestion is contested by such NGOs as Global

Witness, Save the Children, TI, HRW and the Soros OSI. The British government has promised £1 million in technical assistance to any country piloting a transparency scheme. Allegations concerning corruption, however, will probably not disappear, though the processes of corruption may adapt and take new forms.

A second, related, focus for NGOs has been ethical investment. Some funds and equity interests have taken a critical stance on oil companies. Many individuals and institutions are not prepared to rely solely on the due diligence provided by the supposed guardians of executive morality (for example, auditors, accountants, lawyers, analysts, bankers, independent directors and even government officials). They call for the application of a number of tougher ethical standards – and they vote with their money. Ethical disinvestment, which has long been a component of the oil equity markets, has become more prevalent since the Enron debacle.

The Investor Responsibility Research Center in Washington markets a Global Security Risk Monitor database derived in part from work with the Conflict Securities Advisory Group linking firms and assets to "problem countries". This is a growing list. The 260 companies "named" so far are predominantly in the energy industry. This represents a reputational risk for corporate oil that could be reflected to some extent in market behaviour. It was notable, for example that PetroChina encountered difficulties with its IPO in New York. The concern is that large institutional investors (for example, the State of California Pension Fund and members of the National Association of State Auditors, Comptrollers and Treasurers) might use such ratings to shape their weightings on investments in the oil industry.

Following 11 September 2001, the benchmarking of companies against the risk of complicity with terrorism (by assessing their interests in states suspected of sponsoring or condoning terrorism) could also in time become an important tool for allocating investment. Shareholder resolutions have been submitted to companies such as COP and Halliburton calling for directors' committees to be established to review such operations with reference to financial and reputation risks. The protagonists point to the use of offshore subsidiaries to conduct business. Given that the size of holdings often amounts to billions of dollars, such pressures cannot readily be ignored. Some New York pension funds have sought to penalise companies with investments in states such as Iran and Syria.

Many companies have already responded to ethical concerns by promoting social responsibility schemes. Though such schemes often do some good on a local level, such responses are problematic. It is not simply that the schemes themselves are sometimes the subject of criticism or that their costs affect the bottom line (usually only marginally, it should be said). The real problem is that they are, at some level, bound to fail: that is, oil companies cannot fix the social problems of the world on their own. Rather than continue with this strategy, corporate oil may in time prefer to argue for a stricter separation between private industry and state, not unlike the split in Protestant countries between church and state.

In relation to ethical and socially responsible investment, oil companies are currently caught between a rock and a hard place. On the one hand, as the ethical investment movement grows in size and sophistication, companies are likely to face still greater scrutiny. On the other hand, as portfolio diversifies worldwide and more investment is made in riskier frontier zones, they will find themselves increasingly committed to territories where ethical demands are likely to be harder to satisfy.

A third core focus for NGOs has been the environment. The green movement has diverse concerns and constituents. It is comparable to a river with many tributaries. Some NGOs, for example, seek a World Environment Court at which oil companies could face charges for damaging rainforests, animal habitats, wildlife and large riverine systems (especially deltas). Other NGOs overlap with human rights groups in their concern for human habitat and the rights and interests of indigenous peoples. These include some NGO "super-majors" that engage with a wide range of environmental concerns and have established internationally recognised brands or reputations, including Greenpeace, Friends of the Earth and the WWF.

As well as these large organisations, there is a multitude of smaller and more specialised organisations, some of them very active in the oil game. A notable example is Rainforest Action Network, which with around a dozen staff and a budget of $2 million targeted, among others, Texaco for its exploration activities in the Oriente Basin in Peru. There, it is claimed, the company had damaged both the environment and the livelihood of the indigenous people. It has heavily criticised the World Bank (running, for example, full-page advertisements in the *Interna-*

tional Herald Tribune) on the grounds that its financing of fossil fuels has led to destruction of the rainforest.

Many NGOs in the green world are well funded. Funds often come from bequests and donations. The Foundation for Deep Ecology in California acquired a $90 million endowment from Doug Tomkins, an American philanthropist. It has funded the International Forum on Globalisation. Funding for green issues has also come in part from the taxpayer through state allocations.

Many factors may account for the growth of the green lobby and its oil focus. It is able to draw on the emotional impact of the sometimes apocalyptic visions portrayed; it claims scientific support for its concerns; it offers opportunities to supporters in the form of a sense of moral rectitude; and underlying all this is the beguiling, Rousseau-like vision of a return to natural order. To speculate further on the rise of this phenomenon would, however, take us beyond the compass of this book. But the movement's interaction with the oil industry has grown sharply.

Greenpeace is a long-established NGO that has increasingly engaged with the oil industry. It seeks dialogue with corporate oil's senior management teams and, like the companies themselves, employs research and technical expertise. It has been one of the forces behind the demand for environmental reporting. Greenpeace has developed a global reach. It has a US base that facilitates legal actions in that country. With its offshore capability, it is able (unlike many NGOs) to operate in marine environments (including deepwater zones). It is also better placed than many to monitor developments in the industry's frontier zones.

Examples of Greenpeace campaigns include its challenge to

the opposition of ExxonMobil and other companies to the Kyoto agreement, labelling ExxonMobil as "Climate Enemy No. 1". In Australia it has opposed the Stuart Oil Shale Project. It has challenged Australian Oils and Suncor (Canada) in relation to financial and greenhouse gas issues. It has also argued that the American government in effect subsidises corporate oil by, for example, providing defence in the Persian Gulf and schemes such as the strategic oil reserve. Greenpeace has supported its campaigns with headline-grabbing, television-friendly actions such as the blockading of ExxonMobil's French refinery.

The action of Greenpeace and other environmental NGOs has prompted responses on behalf of the industry in the form of the International Association of Oil & Gas Producers (OGP) and IPIECA report on sustainable development issued at the Johannesburg Earth Summit. This was an attempt to illustrate the progress on environmental stewardship made by the industry. Such attempts are not likely, however, to shift Greenpeace from its pulpit. Greenpeace argues that oil dependency fuels climate chaos and that the only solution is for corporate oil to shift investment from carbon fuels to solar and renewable energy, but no measures of the technology development needs and costs to society seem to be provided.

Friends of the Earth (FOE) is another environmental NGO that has engaged with corporate oil over a range of issues. It has often taken equity stakes in companies so that it can propose shareholder resolutions. It has used these at the AGMs of BP and other companies in relation to its campaign against ANWR oil exploration and production and for promoting investment in renewable energy. FOE has secured support for its campaigns from hundreds of NGOs throughout the world. They include churches, wilderness associations, animal rights groups, anti-debt lobbies, civil forums, research institutes, legal bodies, climate networks, social justice groups, charitable foundations, ecology groups and environmental activist groups.

One of FOE's major campaigns has been against the IBRD's financing of fossil fuels. It opposes such financing on the grounds that, it claims, it penalises the poor most (it is said that they suffer through environmental destruction), jeopardises indigenous communities, has negative impacts on women and reduces biodiversity. Moreover, such financing is said to support corporate oil (which supposedly deals with dictatorships and corrupt regimes and is associated with the abuse of human rights), increase developing-country debt, and generate poor-country dependency. Yet the vote in the market by around 6 billion people around the world has favoured lower-cost carbon fuels.

FOE's campaigns against corporate oil have achieved a high profile. At the Johannesburg Earth Summit, for example, it awarded "Green Oscars" to oil companies (notably BP, Shell and ExxonMobil) for "greenwash" (that is, appearing to support environmental ideas but falling short in practice). Such tactics are designed to impose, in effect, a private form of sanctioning in which, it perhaps hoped, the loss of reputation would affect company strategy.

A fourth common focus for NGOs is human rights. Increasingly, oil companies operating in countries governed by unsavoury regimes run the risk of being charged with complicity over human rights abuses. Amnesty International is a leading NGO in this field. It has global connections and conducts a good deal of high-quality research. Its brand enjoys much trust among the public. Its concern with human rights is global. For example, it publishes a map on which it provides a measure of human rights abuses in dozens of countries. Amnesty generally eschews publicity stunts, preferring to operate in a slick, suave, professional manner. A key emphasis is its campaign to persuade oil companies to assess the social impact of their projects. It has more ability than most NGOs to gain the attention of investors and financial institutions.

The fifth estate comprising the NGOs is complemented by several related forces. There are multiple church and faith groups, competitive fuel interests, the global glitterati and intelligentsia, political parties allied to anti-oil interests and some journalists inside the media that support them. These fluid and sometimes mutually supportive lobbies conflate from time to time to present threats and challenges to hydrocarbon interests.

Churches have engaged with the oil industry on a number of issues, such as investment in Sudan. They are concerned about the impact of operations on Christian communities in the developing oil world and have called for disinvestment of church funds from some oil companies on ethical grounds. The nature of such groups varies: on the one hand there are liberals; on the other there are well-organised evangelical or even fundamentalist groups. The latter have been particularly vocal in cases where oil production is a component in the conflict between religious groups (for example, the perceived Muslim–Christian conflict in Sudan which has been infinitely more complex in character).

Corporate oil also faces mounting commercial opposition, in the form of straightforward competition, sometimes allied to regulation, from competitive corporate challengers to the carbon fuel industry. These include coal and nuclear industries and the well-organised lobbies for solar, renewable and hydrogen-based energy.

There is also opposition to oil from individuals in the form not only of donors and NGOs, but also, notably, of assorted glitterati – individuals with, in effect, their own brand and potential for media coverage. For instance, several Hollywood stars have taken high-profile positions against LNG operations in California, and Governor Arnold Schwarzenegger (although Republican) has shifted ground to support climate-change advocates who are perceived to have an anti-oil agenda.

To this list of opponents we should also add fully fledged political parties – notably many reflecting the Green movement with elected representatives in dozens of countries – which influence a number of international or regional institutions, especially in the developing world (for instance the IBRD and its regional affiliates). Whereas opposition from political parties is often conscious and explicit, the impact of the institutions can be implicit – through policies that tilt financing against oil ventures and buttress anti-competition laws, protectionism and regulation.

Lastly, corporate oil now faces more opposition from within selected international media and from some journalists – both inadvertently, sometimes through a lack of understanding of the scientific, technical and industrial issues involved, and sometimes consciously. For some it may simply help sales, while for corporate oil there is a need to be sensitive to media coverage and to seek to avoid bad press. Professional journalists with a bone to pick can do significant damage that may take time and cost to unwind.

The environment in which corporate oil operates has shifted. It would be unwise to base corporate oil portfolio and management strategy on a notion that the coming decades will reflect a linear continuation of the past paradigm. Different challenges from those of the past will materialise. Indeed, the new slew of environmental and climate-change pressures arising on corporate oil, with a host of NGOs found on this bandwagon, means that a new, powerful lobby has emerged. While oil companies in the upstream world are accustomed to dealing with corporate competitors, national oil companies and governments, now they must engage with a wider set of entities that shape global portfolio and strategy decisions.

The new non-formal entities (some of which are well structured and have established histories, so the epithet is not exact) have come to exert a wider set of influences on companies in the world upstream. They include political pressures for disengagement from contentious target countries and investments (notably Sudan, Iran and Angola). There are also campaigns for the insertion of specific non-corporate rights into decision-making over, and ownership of, hydrocarbon assets (for example, concerning rights for lineage-based landowners). Requirements or pressures are placed on companies to engage in foreign ventures at levels increasingly beyond simply their commercial investments, through corporate social responsibility programmes and allied activities. Certain felt needs are generated inside companies to respond to pressures coming from the societies of both the home base and overseas host countries. External pressures create the need within companies to engage in public relations exercises directed at hostile third parties and political entities that advocate a non-oil world order.

These new players in the oil game include the growing range of NGOs, lobby groups, churches, "dissident" shareholders, financial institutions, ethical investment groups and indigenous communities, as well as often less organised social organisations of diverse types, political interest groups and armed groups hostile to foreign companies. These developments have occurred as the formal structures of legitimacy and order, central authority and the nation state have become contested during the march towards fractured globalisation. These informal groups will proliferate still further. If the paradigm becomes less stable and more anarchic, some may transform mere critique into overt hostility. Oil is a prime target for this animus.

The change in operating modes and philosophies of companies – by way of strategic adjustments to accommodate these influences – is already significant. It is reflected in a plethora of mission statements, statements of business principles, good citizenship declarations and the like. *The Shell Report: People, Planets & Profits* is a leading example of industry advocacy and response to date.

The oil industry has learnt to participate in this global oil debate at many levels. Not all companies, however, are equally engaged. They do not all prioritise issues in the same way. Political debates on petroleum will not take place in a social and historical vacuum. Like politics, they will be unpredictable. Humankind will impose its blueprints onto the facts of history and geopolitics.

Change in world oil is likely to be contentious as new maps emerge and older versions of the nation state are displaced. Repressive uni-ethnic powers, contesting multi-ethnic polities and other political formations may emerge in new shapes to impact on corporate oil. Global oil executives are not always well versed in managing all these political complexities. Most have exhibited a desire for them to go away. They won't.

Clearly the "modern barbarian" groups that confront today's equivalent of corporate oil's edifice of Rome are many and various. Here we might return to our analogy with the Roman Empire faced with the barbarian onslaught, which was complex, built on multiple strategies and took a long time to succeed. It will be recalled that, instead of playing to their own strengths, the Romans brought disaster upon themselves. They greatly underestimated the barbarians. Will corporate oil do the same?

Part 4

CORPORATE OIL AND WESTERN FUTURE

The clash between corporate oil (from the American empire and the Western world generally) and the barbarian oil world is not yet settled, at least its final outcome. It may turn out to be an endless struggle. It is, however, a more difficult battle than ever before, and it has been made more so by corporate oil's need to deal with sophisticated threats from within its own domains, just as in Rome it was not only barbarian armies on the outside that brought down the empire but their strategic assaults combined with a weakened polity.

How long can it be before corporate oil raises its awareness of the many hostile forces arraigned against it and determines to learn more about them? Many companies still seem either unconcerned or myopic about the social world in which they operate. The future will not allow them a position of such luxury. The oil game certainly cannot be kept within the confines of the industry itself. Companies are required to confront not only commercial competition but also contestation on the ground, in ideological debate, in the forums of public opinion and in the law courts.

NGOs have grown in volume and in reach within the oil game. They are practised at working in anti-oil coalitions. Corporate oil may ignore them but at some peril. It will in future probably need to employ a variety of new strategies – countering the arguments and ideologies of some, co-opting others, and maximising their distance from those with utopian or fundamentalist agendas aimed at the elimination of the hydrocarbons industry. At present the industry as a whole is weakly informed about such entities – their composition, agenda and strategies.

In its endeavour to penetrate today's hostile domains and operate within them, corporate oil has used many non-commercial strategies not directly related to oil exploration, extraction, refining and marketing. In future, many of these strategies, already deficient, may prove non-viable. New approaches may soon be sought.

Much public relations effort in the oil game has been devoted to the "soft" sell. This is unlikely to prove sufficient to deal with the sharper issues that are likely to arise in future. Oil companies will need to argue their worth more persuasively, basing their case on demonstrating the vital economic roles that their ventures play in the regions in which they operate.

The WSSD's commitment to renewables (though without enforceable targets or timelines) is the harbinger of a wider battle over the legitimacy of the hydrocarbons age. This is reflected already by some hedging in portfolio by, for example, Shell and BP. Many of corporate oil's committed enemies, however, wish to eliminate the oil game as far as possible. This they seek to achieve by raising the above-ground risks, taxes and the costs of continuity, thereby creating a competitive advantage for renewable energy that, they hope, will yield a new global energy order. Increased regulation, mandated policy decisions, oil-related taxes and negative public reaction are among the instruments that will feature in this struggle.

Corporate oil cannot be a substitute for sovereign government. It is unable to meet the vast demands of poverty alleviation. Companies may reconsider such programmes to ensure that these become the task of governments. To the extent that corporate oil has displaced or replaced governments, it risks making itself the target of new hostilities. Oil's social investment efforts often succeed only in creating new divisions. Perhaps oil companies will find strategies for exiting from these self-elected obligations; or else they may bear the consequences of government failures.

Corporate oil will have no choice but to continue to operate in risky environments, especially in the more complex developing world. The legal and other risks involved will necessitate great caution. In hostile environments, companies might decide to reduce the size of the targets that they present. There will be spin-offs, sales of subsidiary interests to locals and possibly even de-consolidation.

In some locations corporate oil has been at risk for many years from hostile, sometimes armed groups. Often they seek to use the industry as a bargaining chip in wider social or political conflicts, as in the Niger Delta. Companies caught in the middle of conflicts may find themselves being forced to choose between sides, thereby aggravating tensions. Often there is no "conflict-lite" solution available. Increasingly, corporate oil will find itself

faced with a choice of protecting itself or abandoning its proper-
ties and leaving. Companies may even need to consider defending
themselves by collaborating with each other in order to match the
coalitions ranged against them, but even this may be restrained by
antitrust legislation for American companies.

Corporate oil understandably places a premium on "stability
for business". Yet many of the regimes that provide most
"stability" – feudal oligarchy in Saudi Arabia, for example – are
of a type that will rarely be condoned in the liberal democracies
in which corporate oil and its main shareholders are domiciled.
Corporate oil must, however, be less interested in global enlightenment than in the business of producing abundant, low-cost
oil. This is the raison d'être for continued corporate existence.
Amid all the discussion of "stakeholders" and "corporate social
responsibility" and "triple bottom lines", there is a risk that some
executives will be seduced by their own corporate spin.

Oil companies, as symbols of corporate wealth (more so than
other resources companies), breed or have inflicted upon them
their own antithesis. Inequality of income and, especially, of
wealth is often at its deepest (or perceived to be so) within certain
petro-states. The sight of islands of luxury amid seas of deprivation breeds deep resentment. In chronically impoverished areas
(such as the Niger Delta) the targeting of corporate oil through
systematic kidnapping, bribery and assaults on oil facilities is an
increasingly "popular" and lucrative option – as, for example,
Chevron has discovered in Escravos.

Corporate oil breeds resentment too as a symbol of a faulted
imperialism – especially, but not exclusively, that of American

empire. American interests encounter opposition not only in economic and political terms, but also in cultural domains as local societies resist homogenisation and the deification of materialism – despite the near-universal presence of McDonald's, American TV, iconography and so on. Few distinctions are observed here: non-US Western oil companies are also targeted as "outposts of empire".

Although globalisation should, according to rationalist models, produce widespread benefits, its praxis has been flawed. Liberal democracy has not triumphed across the world as some "end of history", although the elites of several petro-states like to present a facade of democracy. In any case, the system often fails in practice to produce the comforting neoclassical equilibria that lie at the heart of liberal thinking. Equally, although the process of globalisation may be influenced by international agencies, there is no effective "central machine" available to regulate and manage globalisation's impacts, notwithstanding a plethora of multilateral institutions and the mother of all, the UN. Nor could there be: history teaches that Gosplan-style approaches are likely to fail. The Bretton Woods-style institutions, including the UN, have also shown themselves not to be potent enough for this vast, deeply conflicted task.

In its quest for a strategy with which to survive in these conditions, corporate oil will require sound geopolitical perspectives. This means, as we have seen, an appreciation of the role of empire including, still first and foremost, the American empire but also the new challenges from competitive empires now in ascendance. Though the interests of empire and corporate oil do not always coincide, the latter must at least now take account of the former's strategy and realpolitik, as it has done before in history, as well as the drivers shaping the emerging empires of oil.

Despite the yet-to-be exploited riches in the deepwater Gulf of Mexico (estimated to hold 40 BBOE), US proven/probable reserves amount to no more than 4–5% of the world total. Despite unconventional oil options and Canadian oil sands, America is therefore required to pursue its quest for foreign oil. This global search has taken on a new assertiveness, not least because of the increasing competition for oil from China and the need to hedge as far as possible against its current level of dependency on Middle East suppliers.

There remain a number of possible solutions to the challenge that America and the West face, though none is problem-free. Access for corporate oil in the Middle East would help fulfil supply requirements at, perhaps, softer prices – but that is not in America's gift. Access to Iraq could provide the single biggest counter to dependency on Saudi Arabia (now only at around 10% of imports) – but that has yet to be achieved. Iran appears likely to prove a persistent thorn in the imperial side. Africa, especially the Gulf of Guinea, from Nigeria to Angola, offers large potential that is already being developed – but there is the emerging risk of competitive offtake from China and others. Latin America, too, offers prospects – but there are large political and other obstacles in the way of full development in (and now access to) oil-rich Venezuela. In Central Asia, America faces stiff competition from Russia, China and Asian state oil companies. In theory, Russia's own industry offers an alternative, non-OPEC, source of crude, at least in supplying world markets, more so if the Russian Far East is fully developed – but Russia has its own nationalist hydrocarbons agenda, within which America and Europe have become more supplicant.

For corporate oil there appear to be few easy answers to the need for more resource access and cheap reserves, at least sufficient to increase organic growth in corporate production. Many in Big Oil have underexplored in the past and may pay a price for this neglect in future. New exploration is already under way inside corporate oil, but this process takes much time to yield abundant fruit. Is corporate oil, and with it the foundations of the Western world, at systemic risk as a result of this coalescence of fraught conditions now found in the global oil landscape?

I 2

Clash of cultures in world oil's future

If Machiavelli could survey the shifting world oil game, what finally would that sage of statecraft find? It would be that a new paradigm with resource nationalism, neither yet universal nor unique in history, has emerged in which an apparent irrevocable squeeze is being applied to corporate oil; that companies will need to find new strategies to survive the longer term; that many can and will probably do so, since we are not yet witnessing the ultimate end of corporate oil. Meanwhile, the march towards an end-game, in the form of continuous reshaping, carries on, even if it will be an exceedingly long one that will probably span the century. It is likely, however, that this future trajectory will be uneven at best. Only by managing greater turbulence will corporate oil survive and succeed, and so with it the empires of the West built on hydrocarbons.

Oil claims will lie at the heart of the clash of cultures over natural resources in the present century. A more ruthless struggle for resource control, between states and involving companies, may be anticipated (as is now occurring in the Niger Delta). What would Nigeria look like if, as a result of its deep inner tensions, it fell apart? The conflict would have a heavy impact on oil players there. This may not be the only such case in the future. What would happen if the Middle East oil abundance was somehow compromised?

Nation states may not be able to remain in full control of all future conflicts. There are new "iron curtains", religious walls and ethnic boundaries to consider. They will potentially impose increasingly on societies, political debate and the portfolio interests of global oil players. Each restriction of whatever kind makes for more difficulty in corporate oil access. If a fractured world order deepens, some countries in the second world, outside or omitted from the first world, may even slip into the third world. Despite the uplift of many from measured absolute poverty, a fourth world of grinding global poverty appears destined to expand in absolute terms. This will have ramifications for commercial oil and gas interests which corporate social investment will not be able to remedy.

A number of prospective oil environments may never develop the hard-wired infrastructure necessary as a basis for social stability and secular economic growth. Some countries may exhaust or squander the financial gains from their oil and gas resources without providing for their post-petroleum futures. Even now, many with adverse long-term prospects continue to fail at this critical task. Broader elements of structured chaos may well emerge in the above-ground oil landscape, causing new concerns for companies and in world oil debates.

Oil states with poor track records carry their history with

them, a latent reminder of their failures and a potential source of resurgent crisis. In many current regimes, ageing authoritarian leaders encounter the difficulties of managing transition. Some states that are already "criminalised", or others with malign influences established inside government, may not conform in future to acceptable models for Western oil investment. Purchased politicians, middlemen and apparatchiks will not disappear from future oil landscapes.

Military regimes may attempt to conceal their true colours and present a civilian front. The global elite will want civilisation and its norms to prevail against the tide of problematic regimes, some of which may then be sanctioned. Some states with "murky politics" will nonetheless remain pivotal in world oil and gas. Hence companies will need to do business with them – notwithstanding endemic Western views that somehow human progress should be inexorable or conform to Western models.

Globalisation carries uncalculated risks, and the erosion of the nation state may be one of them, potentially leading to deeper divides in human societies. Some countries, maybe constituting a significant part of the African continent, may come to rest beyond the edge of the developed world as marginalised and segmented entities. Anarchy will not be easy to counter in such circumstances. It has not been readily dealt with in Somalia and the Democratic Republic of Congo, for instance. Some oil players may find themselves at the centre of many of the intense political disputes involved, while others will deliberately seek to avoid presence in such domains.

Powerful historical forces (such as communism) that shaped

the 20th century have flawed legacies that have left their mark on the structures, organisations and body politic of the 21st century. They might blend with new orthodoxies in the form of old and new radical states. Already anti-Western coalitions led by Venezuela have transformed the North Andean region. Russia is set to take on energy superpower status. The contest for control of the Middle East's oil and energy future, long at issue, is at a new advanced stage and conflict there could continue for many years. All this will carry new complexity to international relations, world power balances and the connections with governments which oil companies must negotiate.

The dominant position of Western civilisation – the present driver behind globalisation – should not be assumed to continue, when large swathes of the world contest this model, challenging its precepts and modi operandi. States may only accept globalisation when it suits them. Though this appears to be the case in the oil world now, a transition to autarchy could easily intensify. Many states retain a history of mistrust of the West and antipathy towards one another. Shifting unstable alliances are likely in this newly fractured environment.

Dynastic states, much in evidence today (the DRC is one case, also Brunei, Jordan, Saudi Arabia and the UAE), may in time face challenges, with their survival placed at risk. In the coming quarter of a century there will be uncertain transitions in Libya, Nigeria, Venezuela, Indonesia, Angola and other key zones in which significant oil reserves are found. The post-transitional outcomes may not always easily be aligned to the interests of democratic capitalism and corporate oil.

Some states may remain centrally controlled, not succumbing to the march of the markets and the virtues of energy deregulation, liberalisation and privatisation as proclaimed by the EU and its culturally conditioned oil companies. Fundamentalism in many petroliferous states has undoubted potential to grow (not only in the Islamic world), so challenging most secular orders in its path, including top-down controlled regimes in Maghreb and Arab societies. The future of Africa, like its past, is unlikely to be free of turbulence.

Globalisation will bring with it more interconnections of random determination: isolated events will have little chance of remaining so. The geological and market tapestry of world oil and gas could be affected more rapidly and profoundly than before by such developments. Warlords in one area or fratricidal strife in another might have impacts on companies elsewhere. History preaches the instability of steady-state systems over longish periods.

Some states built on weak ideological foundations may flounder, as their founding mythologies fossilise and are replaced by new images and revisionist histories. Autocrats may seek to apply Western democratic principles in their own interests, as has been found in some African and Asian states. Police states will not fold and disappear because "world opinion" deems that best. The infrastructure and political architecture of much that might be problematic with the world will not erode naturally. The appetite on the world stage to fix all problems appears weak and may be diminishing. Indeed, many tragedies in 20th-century world politics have not been washed away by UN resolutions and temporary truces. Their reappearance in new forms should not be discounted. In some modernising states, political decay and recidivist lapses may yet be encountered, with unexpected consequences. The heavy incidence of wars, and their complex relationships with the oil game, may continue.

"Big Power" interests and relationships reflecting competitive empires (involving America, Europe, Russia, China and India) may alter, ethnic cleansing reappear or repeat, oligarchies be displaced, tribal and clan rivalries emerge, and warlords take on older and crumbling political powers – all to the detriment of some corporate oil interests. Presumptions within corporate oil of a benign future should be discounted. By 2030 a new set of local heirs to the global political order and its oil reserves will walk upon the world energy stage. Corporate oil may have little connection with the inheritors of the future and may even be maltreated by them.

If the "end of modern history" is materialism and the consumer culture, and if that is built largely on petroleum, it will emerge in a world of growth and rising oil demand subject to intense competition and dispute. Indeed, if order breaks down amid failures to raise standards for the middle class beyond the old imperial centres (this middle-class growth is evident now in China and India) and for the world's disenfranchised populaces, a return to presidentialism, coups d'état or anarchic revolts might become more common. Corporate oil is not intended, and will not be able, to remedy all these maladies. Petrodollars and those that control them are by most calculations incapable of delivering *la dolce vita* for all, despite many well-intentioned company efforts to ameliorate social disadvantage in the localities where they place their assets.

More companies from the West will need to trek further into the cauldron of uncertainty that makes up the developing world where political earthquakes could surprise corporate oil, given that much of the terrain where they must invest is fragile in terms of institutional order. There they will find more local competitors drawn from the ranks of the barbarians. Alliances with them will proliferate and so dilute corporate oil in some measure.

There oil companies will confront familiar issues, but also new threats: dissidents, sects, opposition parties, ethnic rivals, ideologues, activists, armed militia and mendacious politicians not aligned with the governing elites with whom corporate oil does business. Deals struck with foreign aristocracies and elites abroad will not always be legitimised locally or even be uncontested in the investor's home base. The unwinding of older empires could carry new risks and threats to the establishments that in the past ruled world oil. More nation states on the global stage will lead to a more balkanised world with a greater array of contested legitimacy. Boards and shareholders may require a morality in transactions that cannot be met in some parts of the newly emerging oil world. Self-interest may increasingly conflict with today's new Western penchant for moral absolutism in more turbulent times, thereby greatly narrowing the usual boundaries for acceptable investment compromise.

Exactly how many convulsive "Yugoslavia-style" break-ups might yet emerge in the Middle East, Central Asia, Africa, Asia, Latin America and elsewhere over the coming quarter of a century cannot be predicted, but there may well be some. The underpinnings of their causal origins can be related to many existing global political and economic structures. Ecological pressures and predictable demographics may add to this cocktail of uncertainty. Stabilising forces are often poised in a delicate balance with the forces of instability, as now seen in many African states. Democracy on this continent is in the stages of early experimentation and by no means the inevitable "natural order" of the future. The "heart of darkness" was *terra incognita* and this Conradian paradigm is not restricted to the African predicament.

Unknowns abound in and around many pivotal oil states – Saudi Arabia, Russia, Iraq, Libya, Indonesia, Iran, Colombia and Venezuela. Most Central Asian entities have had but a few years of separation from the clutches, not always wholly removed, of mother Russia. Even internal "quasi-autonomous zones" like Xinjiang in China have demonstrated anti-Beijing irredentist tendencies. The Moscow–Beijing démarche and cohabitation now witnessed may not last for ever.

Globalisation has also brought dollarisation (here a symbol of hard cash availability, despite dollar weakness) across the economic and oil landscape, and with it the unevenness that this process typically produces for groups that lack access to foreign exchange and must hence survive within the straitjackets of diminished local currencies. Such segmentation can readily give rise to powerful instabilities that might dramatically realign politics. The world may yet divide into those with dollars (or other strong currencies) and those without – the latter left to survive on devalued and worthless surrogates or old forms of barter and exchange. Some states (Iran) have already challenged the omnipotence of the dollar in oil transactions.

Remaining "empires" from the past, which still exist as large states, might in future go the way of other imperial powers, with serious and violent transitional consequences for their linguistic, ethnic and regional unities. It is by no means clear that Russia's boundaries have been for ever settled. Nor is it certain that China will remain untouched by internal separatism.

The number of countries falling behind the leaders in globalisation is large and their economic future is uncertain. Some may never achieve integration into the new world order on satisfactory terms, and more could turn their back on this early 21st-century model. This could involve resistance to the corporate oil world, which once took its global remit and a benign world landscape as the future playing field for granted. The Western world and corporate oil in general could decide to concentrate more on "strategic zones" that mean most to their critical oil interests. They may neglect problematic regions. Such focus could flow from self-interest, a key driver in corporate strategies. The shift to non-oil options and technologies in energy may be an early sign of an adjustment within a world system fearful of global warming and its expected or predicted consequences.

Debates on all these issues will be intense, critical to global oil strategy and infused with a wide set of divergent positions that make for "noisy discourse". In this way, the world petroleum stage is now partially globalised but heavily contested. It will become more so in the years ahead. Political debate on petroleum – the most "political" of commodities – as conducted with growing acrimony in many forums (including cyberspace) is likely to sharpen. Companies may need to be prepared to defend their commercial interests and their reputations in such debates.

The future is bringing to the fore a range of apparently non-oil issues with which corporate oil is already forced to engage – transparency requirements are one. A canvass of the issues expected in the coming period of high turbulence, as it might affect corporate oil, illustrates the challenges to come. These issues will produce contentious global debate involving many groups that are hostile to corporate oil. Some of this debate will originate over issues mainly within world oil, as well as some from outside it and others derived from areas that cross the once clear boundaries between the oil world and the rest. In future, there will be little or no place left to hide: a multiplicity of issues will be considered as, or be made, germane to oil, since it is now considered significant to most societies in the world.

The culture of human rights, an evolving yardstick, is deeply embedded in the Western psyche and political practice, and likewise finds local expression in most cultures even if in different forms. If the future world becomes less conducive to normative human rights conditions, so the contestation of corporate oil's presence inside insecure barbarian oil worlds may grow in the West. The number of organisations and NGOs in this sphere has increased and their involvement with oil has grown. Affected companies have been forced into responses. Others can expect similar attention in future. The UK and US governments issued rules on human rights related to armed security in 2001, with the principles supported by Human Rights Watch, Amnesty International, International Alert, Lawyers Committee for Human Rights, Fund for Peace and others. Human rights have been, and are likely to remain, a vexed question in a more turbulent oil world. Corporate oil must now tread with care in this arena. In future, some areas of venture may become off limits or be entered with additional associated risk (as, for example, now applies in Darfur).

Not all armed security at oil installations in new venture environments is provided by the public security forces. In several cases companies have hired private security firms. Companies have responsibility for employee security in their operations. So pervasive is the issue now that Shell has had recourse to armed security in 27 countries, with the company seeking conformity to international guidelines for its operations and contractors. This model appears to be the harbinger of the future. Nor has armed attack on oil players been restricted to land-based conflicts. Piracy is widespread. Worldwide there are about 300 attacks on oil industry shipping each year. This necessitates contingency plans based on risk assessment, deterrence, vigilance and even passive resistance. The threats to the oil industry are generally growing in scale and incidence across wider terrains.

Rogue states have increased in number and longevity. Some of these states hold oil reserves and prospective acreage acquired by the industry. Iraq was seen by many non-formal entities and even governments as such a case (with its invasion having already had huge implications for the global oil industry). The full sanctioning of Iran would likewise cut out some potential from the world oil system. Wider conflict in and around the Gulf, in the

Gulf of Hormuz choke-point, or in Saudi Arabia would be potentially lethal for world oil markets. In general, the more ethical the definitions of acceptable practice applied by corporate oil's puritans, the greater the number of rogue states that will be found. A shrinking universe of opportunity, including from potential nationalisations, may await some in corporate oil as permitted choices become more restricted.

The world is not short of conflict zones, and some involve issues directly related to petroleum resources. Many protagonists, whether states or other forces, seek to involve oil companies on their side or, if unsuccessful in this task, to target them as antagonists. Corporate oil's presumption of neutrality will be questioned more widely in the future.

Sanctioned countries such as Syria, Sudan, Myanmar and Iran already face restrictions on investments in and sales of oil and gas. Many have high fossil fuel potential and opportunities, but a number have repressive regimes and poor track records, as judged by Western interests. Debate on the ethics and efficacy of sanctions is intensifying. Many treat sanctions as a means of foreign policy but they may cause poverty (as was the case in Iraq). Issues such as the ethics of generalised sanctions and the arguments for and against "smart sanctions" have proved contentious. They have also been a further cause of disunity among world oil powers (notably Russia and China in regard to Sudan and Iran). Oil companies are finding themselves within the sights of the different vested interests that revolve around these controversial issues, whereas most have a direct interest in sanctions amelioration and the lifting of *force majeure* constraints. But the

American government and others in the OECD may add to the list of recalcitrant states that are seen to deserve punishment in this manner. Corporate oil will be a loser in such a process as others intervene to take their place.

Over the next quarter of a century, some countries will face oil maturation problems and the need to design post-petroleum strategies. China's industry is already reaching out across the world in order to meet this challenge. Smaller states such as Bahrain may face such a challenge well before 2030. Several states have modest reserves and rising oil production or demand needs. In some places corporate oil may find itself blamed for economic demise following the shrinkage of resources.

The world appears unlikely to move towards a much more even distribution of wealth and income. As history has demonstrated, this is a difficult area in which to design and execute desired change. The aggravated argument over the role of corporate oil in relation to poverty will therefore increase rather than diminish, even if rising per head incomes are achieved.

The terrorist world has now selected the oil milieu as a target for hostilities. This is a global phenomenon that is unlikely to diminish in future as more groups organise to oppose America, its primary and secondary allies, and hence corporate oil.

Corporate oil is already part of the global struggle for control and access to oil-rich zones and to equity supply. It will attract more threats in this task. And it will risk charges of complicity over human rights abuses in some countries simply by virtue of operating there. Corporate oil has come to depend on its social and humanitarian strategies to deflect such criticism, but there is already a growing question mark over whether such initiatives, by nature limited and in effect marginal to the scale of the problems, can be relied upon in the long term. New approaches may be needed, and some companies might accept that this form of "band-aid" serves only to attract even more criticism. Notwithstanding such efforts, corporate oil will perhaps be perceived as a cause and agent of exploitation in the developing world. It will attract hostility for this role regardless of its social projects and public relations spin.

Corporate oil ventures facing direct physical threats may need increasingly to rely on military and security organisations, a practice that will attract greater criticism in and of the West. Many local communities or ethnic groups will consider themselves to be disadvantaged by the actions of the oil industry and some will take the kind of armed action now seen in the Niger Delta. NGOs will take up this cause, lobby against corporate oil and challenge it in the courts. Meanwhile, a new mini-empire in the oil game is flourishing in the creeks and wetlands of the delta.

The actions of the military within praetorian regimes will continue to pose a risk to corporate oil. It may be expected that some problematic events will occur as states seek to protect their oil interests against dissident groups. Oil companies may need armed protection in some countries. This will involve difficult choices. Undoubtedly, corporate oil will be accused of fostering links with military groups and facilitating the supply of armaments.

The interplay of such complex issues and interest groups in and around the matrix of empire and its assault by barbarian worlds is reshaping the global oil game in new and sometimes unexpected ways. Corporate oil will have to become fully aware of these new complexities – and the widening set of non-formal influences – which nowadays and in future will have a growing impact on the world petroleum industry. It will be forced equally to adjust to the facts of shifting empires in oil that are now redesigning the global oil terrain.

It may be seen from the shifting sands that the American oil empire is diminishing in importance, along with Western variants, at least in terms of control over conventional oil if not yet all hydrocarbons. New empires of oil have arrived on the world landscape, and the conditions of this new oil game are not propitious for the survival and development of corporate oil on the well-worn pathways taken in the past. The paradigm shift appears permanent. Our hydrocarbon world has altered in several fundamental ways, as it has in the past. Corporate oil survived many past transformations: it can do so again – but it will only do so inside a new oil world.

There will, therefore, be no single empire of oil in the 21st century. The unwritten history of the future will be formed by the dance of many oil empires as they jockey for position and mutate in form. The barbarians have long since arrived at the gates of world oil, have achieved control of many commanding heights and should not now be expected to retreat.

Notes

1 See Duncan Clarke, *The Battle for Barrels: Peak Oil Myths and World Oil Futures*, Profile Books, 2007.

2 Michael White, *Machiavelli: A Man Misunderstood*, Little, Brown, 2004.

3 Niccolò Machiavelli, *The Prince*, translated by George Bull, Penguin Books, 2004.

4 See Robert D. Kaplan, *Warrior Politics*, Random House, 2003, chapter 5.

5 Edward Gibbon, *The Decline and Fall of the Roman Empire*, Alfred A. Knopf, 1993 (volumes 1, 2 and 3).

6 Peter R. Odell, *Oil and World Power*, Penguin, 1971 (1st edition), with subsequently seven editions to 1986. Odell has been a prolific writer on world oil affairs for over forty years (see for instance Odell, *The Economic Geography of Oil*, G. Bell and Sons, London, 1964), and his most recent work, a landmark evaluation, sums up much of this economic thinking and new optics, as found in Odell, *Why Carbon Fuels Will Dominate the 21st Century's Global Energy Economy*, Multi-Science Publishing Co. Ltd, 2004. In this significant analytical review of world energy can be found references to most of Odell's earlier writings as well as important contributions from others.

7   Daniel Yergin, *The Prize: The Epic Quest for Oil, Money and Power*, Touchstone, 1991.

8   See Clarke, op. cit.

9   Robert Kagan, *Paradise and Power: America and Europe in the New World Order*, Atlantic Books, 2003.

10 Robert Cooper, *The Breaking of Nations: Order and Chaos in the Twenty-first Century*, Atlantic Books, 2003.

11 Robert D. Kaplan, *The Coming Anarchy, Shattering the Dreams of the Post Cold War*, Vintage Books, 2000.

12 Odell, *Why Carbon Fuels Will Dominate the 21st Century's Global Energy Economy*, op. cit.

13 Ibid. Note the bibliography that lists the prolific research and publications on the oil industry undertaken by Odell over a period of some forty years.

14 Leonardo Maugeri, *The Age of Oil: The Mythology, History, and Future of the World's Most Controversial Resource*, Praeger, 2006.

15 Ibid., p. xi.

16 Ibid., p. xvi.

17 For a summary of some thinking on this set of ideas, see Eric D. Beinhocker, *The Origin of Wealth: Evolution, Complexity, and the Radical Remaking of Economics*, Random House, 2006.

18 Michael T. Klare, *Resource Wars: The New Landscape of Global Conflict*, Metropolitan Books, 2001.

19 Ibid., p. 29.

20 Stephen Pelletiere, *America's Oil Wars*, Praeger, 2004.

21 Rashid Khalidi, *Resurrecting Empire: Western Footprints and America's Perilous Path in the Middle East*, Beacon Press, 2004.

22 Lutz Kleveman, *The New Great Game: Blood and Oil in Central Asia*, Atlantic Books, 2003. See also Rosemarie Forsythe, *The Politics of Oil in the Caucasus and Central Asia*, Adelphi Paper 300, Oxford University Press, 1996.

23 Kleveman, op. cit., p. 258.

24 See variously: Andy Stern, *Who Won the Oil Wars? Why Governments Wage War for Oil Rights*, Conspiracy Books, 2005; Garry Leech, *Crude Interventions: The United States, Oil and the New World (Dis)Order*, Zed Books, 2006; William Engdahl, *A Century of War: Anglo-American Oil Politics and the New World Order*, Pluto Press, 2004. On American oil thirst and its consequences, in a critique of this issue, with some hard words for the corporate oil players, see Linda McQuaig, *It's the Crude Dude: Greed, Gas, War, and the American Way*, Thomas Dunne Books, 2004.

25 Engdahl, op. cit., p. ix.

26 Ibid., pp. 263–4.

27 Some of Engdahl's views border on the conspiratorial. The so-called secret deliberations of the Bilderberg Group and Trilateral Commission, among others, including Henry Kissinger, are invested with enormous power to influence and direct control over world oil affairs. This would amuse many players in the global industry who find that their own ability to influence let alone control governments, markets,

upstream conditions and the industry are a constant challenge.

28 Stern, op. cit.
29 Ibid., p. 7.
30 Ibid., p. 8.
31 Ibid., p. 220.
32 Ibid., p. 246; this is the concluding remark of this book.
33 Leech, op. cit.
34 Ibid., p. 7.
35 Leech, op. cit, p. 222; this is the concluding citation of this book.
36 Andy Rowell, James Marriott & Lorne Stockman, *The Next Gulf: London, Washington and Oil Conflict in Nigeria*, Constable, 2005. See also, on Nigerian oil issues, Kenneth Omeje, *High Stakes and Stakeholders: Oil Conflict and Security in Nigeria*, Ashgate, 2006, where the multiple complexities of oil wars in the Delta are exposed and related to the rentier state, corruption, security, environment, instability, ethnicities, local grievances and more besides. A conventional prescriptive view is given in Human Rights Watch, *The Price of Oil: Corporate Responsibility and Human Rights Violations in Nigeria's Oil Producing Communities*, Human Rights Watch, 1999. And on the thesis of corporate culpability, see Ike Okonta and Oronto Douglas, *Where Vultures Feast: Shell, Human Rights and Oil in the Niger Delta*, Verso, 2003, with Shell seen as a colonial force and crude mogul in the oil game. Wider critical views on Shell and its various crises, notably

oil reserves, governance, management and public relations issues, are found in Ian Cummins and John Beasant, *Shell Shock: The Secrets and Spin of an Oil Giant*, Mainstream Publishing, 2005.

37 Ibid., p. 188.

38 See here Raymond J. Learsey, *Over A Barrel: Breaking the Middle East Oil Cartel*, Nelson Current, 2005.

39 Among key works to note, see Zbigniew Brzezinski, *The Grand Chessboard: American Primacy and its Geostrategic Imperatives*, Basic Books, 1997; Rashid Khalidi, *Resurrecting Empire: Western Footprints and America's Perilous Path in the Middle East*, op. cit.; Robert Kagan, *Paradise and Power: America and Europe in the New World Order*, op. cit.; Robert Cooper, *The Breaking of Nations: Order and Chaos in the Twenty-first Century*, op. cit.; Robert D. Kaplan, *The Coming Anarchy: Shattering the Dreams of the Post Cold War*, op. cit. (plus several related books by Kaplan); Roger Scruton, *The West and the Rest: Globalisation and the Terrorist Threat*, Continuum, 2002.

40 Peter Heather, *The Fall of the Roman Empire: A New History*, Macmillan, 2005. The quotations that follow in this paragraph are from p. xi.

41 See Dominique Moisi, "Reinventing the West", *Foreign Affairs*, November/December 2003.

42 Richard Rudgley, *Barbarians: Secrets of the Dark Ages*, Channel 4 Books, 2002.

43 See Duncan G. Clarke, *Domestic Workers in Rhodesia: The Economics of Masters and Servants*, Mambo Press, Gwelo, 1974, for a study of the theory and practice of these relationships.

44 In respect of Africa, see Martin Meredith, *The State of Africa: A History of Fifty Years of Independence*, The Free Press, 2005.

45 Niall Ferguson, *Colossus: The Rise and Fall of the American Empire*, Penguin Books, 2005.

46 It should be made clear that all references to NGOs here in regard to hostility or adversarial relationship to hydrocarbons and corporate oil are meant to focus on only such NGOs and not to implicate all NGOs, since many function in co-operation with oil companies. At the same time, this is a complex arena and some with much legitimacy and credibility may at times clash with oil interests on selected causes and issues.

47 Ferguson, op. cit., p. 29.

48 Niall Ferguson, "Empires with Expiration Dates", *Foreign Policy*, September/October, 2006.

49 Niall Ferguson, "The Next War of the World", *Foreign Affairs*, September/October, 2006.

50 See Samuel Huntington, *The Clash of Civilizations and the Remaking of World Order*, Simon & Schuster, 1996.

51 See Brzezinski, op. cit.

52 See "Pentagon considers separate US command for Africa", *International Herald Tribune*, 22 December 2006.

53 See Brzezinski, op. cit., p. 215.

54 Francis Fukuyama, *The End of History and the Last Man*, Free Press, 1992.

55 See Nicholas Shaxson, *Poisoned Wells: The Dirty Politics of African Oil*, Palgrave Macmillan, 2007. The bulk of this otherwise informative book is a critique of individuals within the political oil elites in selected countries, and the view taken on the oil curse in effect excuses oil companies and states from any causal connection to the oil curse, so depicted. While novel as an idea, it is not persuasive, while the notion that governments (including policies that they practise) have no liability or impact in their own backyard is bizarre.

56 For the many complexities around this long-drawn-out conflict, see Nicholas Coghlan, *Far in the Waste Sudan: On Assignment in Africa*, McGill-Queen's University Press, 2005.

57 See "The World in 2007: A pause in democracy's march", *The Economist*, 2006.

58 For some views on the energy games of China, Iran and Russia, see Kleveman, op. cit., and on China's oil initiatives, see George Orwel, *Black Gold: The New Frontiers in Oil for Investors*, Wiley, 2006, pp. 130–5. American strategy in oil is treated in Maugeri, op. cit.

59 See here Chalmers Johnson, "Republic or Empire", *Harper's Magazine*, January 2007.

60 IEA, *World Energy Outlook*, Paris, 2006.

61 Chongololos is a concept used by Global Pacific & Partners in Africa (applicable elsewhere) to describe the flood of

new small players, on the AIM or from private capital markets, that have entered the upstream. The chongololo is an African millipede, common in Southern and Central Africa, which comes out after the rains. The phenomenon of small-company proliferation is not restricted to Africa and is now evident worldwide.

62 Drawn from Strategy Briefings conducted in and on Africa, Asia, Latin America, Middle East, national oil companies worldwide, and on world upstream strategy, with presentations made by the author for Global Pacific & Partners (for details of the firm's Strategy Briefings, see www.petro21.com).

63 See here Richard N. Hass, "The New Middle East", *Foreign Affairs*, November/December, 2006.

64 Roger Howard, *Iran Oil: The New Middle East Challenge to America*, I.B. Tauris & Co Ltd, London, 2007.

65 For views on Venezuela critical of the critics of Chavez, and the disputes within the country over oil, see Nikolas Kozloff, *Hugo Chavez: Oil, Politics, and the Challenge to the US*, Palgrave, 2006.

66 On the original scramble and its rich and complex history, see the classic historical work by Thomas Pakenham, *The Scramble For Africa*, Jonathan Ball Publishers, 1991. In the case of African hydrocarbons, there is an analogous report published by Global Pacific & Partners (Duncan Clarke, "The Third Scramble for Africa: Origins, Insights & Dynamics To 2025: The History & Future of African Petroleum", 2003) which was released to private clients.

The author conducts *The Scramble for Africa: Strategy Briefing* each year in Cape Town at the Africa Upstream Conference, focusing on the oil and gas game across the continent.

67 See Clarke, *The Battle for Barrels*, op. cit., pp.168–76.

68 Michael Holman, "Foreign Aid II: This kind of "help" is just no help at all", *The Africa Report*, No. 4, October 2006.

69 See "Boots on the ground", *Africa Confidential*, 3 November 2006, Vol. 47, No. 22. See also "Scramble for African Oil", *New African*, July 2006.

70 See Ricardo Soares de Olivera, "The Geopolitics of Chinese Oil Investments in Africa", paper delivered at China–Africa Relations Conference, Sidney Sussex College, Cambridge, 12–13 July 2006.

71 Much is reported on Nigeria's oil corruption: here see some details in A. A. Mwankwo, *Nigeria: The Stolen Billions*, Fourth Dimension Publishing Ltd, 1999.

72 For an insight on this history and the complexities see Coghlan, op. cit.

73 Jean-Marc Balencie and Arnaud de La Grange, *Mondes Rebelles: Guerilla, Milices, Groupes Terroristes*, Editions Michalon, 2001.

74 For some views on this and implications for the industry, as well as multiple risks in the global oil architecture, see Neal Adams, *Terrorism and Oil*, Penn Well Corporation, 2003.

75 When at Petroconsultants in the mid-1980s, the author developed with colleagues the first worldwide Country Petroleum Risk Environment, a hydrocarbon-specific

evaluation of political and commercial risks in oil and gas across 90 countries. This became the subject of an annual report that later was modified and put into quantitative format and still serves as a key instrument in the evaluations provided to the industry by IHS Energy (which acquired Petroconsultants).

76 See numerous corporate oil websites for such annual reports, updated now on environment and social policy. This has become a veritable mini-industry within the world of oil and now absorbs much time, cost and effort on the part of the companies. Contrary views abound about corporate shortcomings and alleged malfeasance, such as Suzana Sawyer, *Crude Chronicles: Indigenous Politics, Multinational Oil, and Neoliberalism in Ecuador*, Duke University Press, 2004.

77 See Francois-Xavier Verschave, *Noir silence: Qui arrêtera la Francafrique?*, Editions des Arènes, 2000.

78 Anti-oil initiatives arise within many events, often web-marketed. Some take the form of set-piece gatherings and sometimes organised confrontations that have taken on a global calendar at scheduled world events, to date in Seattle, Prague, Washington, Nice, Quebec City, Genoa, Vienna, Windsor (Ontario), Millau (France), Okinawa, Philadelphia, Los Angeles, Melbourne, Seoul, The Hague, Davos, Kananaskis (Canada), London and elsewhere. The list will grow as the G8, IMF, WTO and multilateral organisations become specific targets of the deconstructionists of globalisation.

79 See Susan George, "Corporate Globalisation", in Emma
    Bircham and John Charlton (eds), *Anti Capitalism: A Guide
    to the Movement*, Bookmark Publications, 2001.

80 Critique of Shell's public relations and their corporate
    social responsibility programmes in Nigeria and the Delta
    may be found in Rowell *et al.*, op. cit. See also Leech, op.
    cit., which also treats with conflict and similar issues in
    Colombia.

81 Bjorn Lomborg, *The Skeptical Environmentalist*,
    Cambridge University Press, 2001. See also Lomborg's
    website: www.lomborg.com

82 For insights into a selection of the various ideologies
    and programmes, see Bircham and Charlton (eds), op.
    cit., where essays by Susan George and other prominent
    adherents of this movement may be found.

83 See Steve A. Yetiv, *Crude Awakenings: Global Oil Security
    and American Foreign Policy*, Cornell University Press,
    2004.

84 For some idea of the entities involved and issues
    canvassed, see Project Underground and www.moles.org/
    projectunderground and www.globalpolicy.org for further
    references, as well as CorpWatch at www.corpwatch.org
    where many protests against oil and companies are noted.
    Also see the list of organisations that support the Publish
    What You Pay initiative on www.publishwhatyoupay.org;
    and for documents on the anti-oil lobbies see Oil Watch
    on www.oilwatch.org where several worldwide anti-oil
    campaigns are cited.

Appendix

Abbreviations and measures

Abbreviations

AFL-CIO	American Federation of Labour and Congress of Industrial Organizations
AGM	annual general meeting
ANC	African National Congress
ANWR	Alaska National Wildlife Refuge
ASEAN	Association of South-East Asian Nations
CEO	chief executive officer
CIA	Central Intelligence Agency
DRC	Democratic Republic of Congo
EBRD	European Bank for Reconstruction and Development
EEZ	exclusive economic zone
EITI	Extractive Industries Transparency Initiative
EOR	enhanced oil recovery
EU	European Union
FARC	Fuerzas Armadas Revolucionarias de Colombia (Revolutionary Armed Forces of Colombia)

FOE	Friends of the Earth
FPSO	floating production storage and offloading
GDP	gross domestic product
GTL	gas-to-liquids
HRW	Human Rights Watch
IBRD	International Bank for Reconstruction and Development
ICC	International Chamber of Commerce
IEA	International Energy Agency
IFC	International Finance Corporation
IMF	International Monetary Fund
IPIECA	International Petroleum Industry Environmental Conservation Association
IPO	Initial Public Offering
JV	joint venture
LNG	liquefied natural gas
MEND	Movement for the Emancipation of the Niger Delta
MOU	memorandum of understanding
NATO	North Atlantic Treaty Organisation
NGO	non-governmental organisation
OECD	Organisation of Economic Co-operation and Development
OGP	International Association of Oil & Gas Producers
OPEC	Organisation of Petroleum Exporting Countries
OSI	Open Society Initiative
SADR	Saharawi Arab Democratic Republic
SPLA	Sudanese People's Liberation Army
SPLM	Sudanese People's Liberation Movement

TFG	transitional federal government (Somalia)
TI	Transparency International
UAE	United Arab Emirates
UN	United Nations
UNICEF	United Nations Children's Fund
WSSD	World Summit on Sustainable Development
WWF	World Wildlife Fund

Selected companies

BP	British Petroleum
CEF	Central Energy Fund (South Africa)
CNOOC	Chinese National Offshore Oil Corporation
CNPC	Chinese National Petroleum Corporation
COP	ConocoPhillips
DNO	Det Norske Oljeselskap
GNPOC	Greater Nile Petroleum Operating Company
ILSA	Iran-Libya Sanctions Act (US)
INOC	Iraq National Oil Corporation
IOC	Indian Oil Corporation
Jogmec	Japan Oil, Gas & Metals National Corporation
KNOC	Korea National Oil Corporation
KPC	Kuwait Petroleum Company
MOGE	Myanmar Oil & Gas Enterprise
NIOC	National Iranian Oil Corporation
NNPC	Nigerian National Petroleum Corporation
NOC	National Oil Corporation (Libya)

OGDC	Oil & Gas Development Corporation (Pakistan)
OIL	Oil India Limited
ONGC	Oil & Natural Gas Corporation (India)
PDVSA	Petroleos de Venezuela SA
TPAO	Turkish Petroleum Corporation
UTP	Union Texas Petroleum
YPFB	Yacimientos Petroliferos Fiscales Bolivianos

Measures

BBLS	billion barrels
BLS	barrel (oil)
BOE	barrel of oil equivalent
MBOPD	thousand barrels of oil per day
MMBLS	million barrels
MMBOPD	million barrels of oil per day
TBLS	trillion barrels
TCF	trillion cubic feet

Index